T0367908

Floods in a Changing Climate

Inundation Modelling

Floodplains are among the most valuable ecosystems for supporting biodiversity and providing environmental services, and are also home to approximately one-sixth of the world population. As a result, flood disasters currently affect more than 100 million people a year. Flood inundation models are a valuable tool in mitigating increasing flood fatalities and losses. This book demonstrates how these models enable us to make hazard predictions for floodplains, support appropriate land use and urban planning, and help discourage new human settlements in flood-prone areas. It provides an understanding of hydraulic modelling and floodplain dynamics, with a key focus on state-of-the-art remote sensing data, and methods to estimate and communicate uncertainty. Additional software and data tools to support the book are accessible online at www.cambridge.org/dibaldassarre.

This is an important resource for academic researchers in the fields of hydrology, climate change, environmental science and natural hazards, and will also be invaluable to professionals and policy-makers working in flood risk mitigation, hydraulic engineering and remote sensing.

This volume is the third in a collection of four books within the International Hydrology Series on flood disaster management theory and practice within the context of anthropogenic climate change. The other books are:

1 – Floods in a Changing Climate: Extreme Precipitation *by Ramesh Teegavarapu*
2 – Floods in a Changing Climate: Hydrologic Modeling *by P. P. Mujumdar and D. Nagesh Kumar*
4 – Floods in a Changing Climate: Risk Management *by Slodoban Simonović*

GIULIANO DI BALDASSARRE is a Senior Lecturer at the UNESCO-IHE Institute for Water Education in Delft, the Netherlands, and also works as the Coordinator of the EC FP7 KUL-TURisk project, which aims at prevention of water-related disasters through evaluation of different risk prevention measures. His teaching and research interests include: floodplain processes and inundation modelling, hydroinformatics and remote sensing data, statistical hydrology, and flood management under uncertainty. Dr Di Baldassarre serves as an editor of *Hydrology and Earth System Sciences*, and guest editor of *Physics and Chemistry of the Earth and Hydrological Processes*, and is also the author of more than 30 peer-reviewed journal articles and more than 60 other publications including invited book chapters and conference proceedings. His scientific papers have received more than 200 citations, and his achievements have been recognized by a number of prizes, including the Outstanding Young Scientist Award from the European Geosciences Union.

INTERNATIONAL HYDROLOGY SERIES

The **International Hydrological Programme** (IHP) was established by the United Nations Educational, Scientific and Cultural Organization (UNESCO) in 1975 as the successor to the International Hydrological Decade. The long-term goal of the IHP is to advance our understanding of processes occurring in the water cycle and to integrate this knowledge into water resources management. The IHP is the only UN science and educational programme in the field of water resources, and one of its outputs has been a steady stream of technical and information documents aimed at water specialists and decision-makers.

The **International Hydrology Series** has been developed by the IHP in collaboration with Cambridge University Press as a major collection of research monographs, synthesis volumes, and graduate texts on the subject of water. Authoritative and international in scope, the various books within the series all contribute to the aims of the IHP in improving scientific and technical knowledge of fresh-water processes, in providing research know-how and in stimulating the responsible management of water resources.

Floods in a Changing Climate

Inundation Modelling

Giuliano Di Baldassarre

UNESCO-IHE Institute for Water Education

Shaftesbury Road, Cambridge CB2 8EA, United Kingdom

One Liberty Plaza, 20th Floor, New York, NY 10006, USA

477 Williamstown Road, Port Melbourne, VIC 3207, Australia

314–321, 3rd Floor, Plot 3, Splendor Forum, Jasola District Centre, New Delhi – 110025, India

103 Penang Road, #05–06/07, Visioncrest Commercial, Singapore 238467

Cambridge University Press is part of Cambridge University Press & Assessment,
a department of the University of Cambridge.

We share the University's mission to contribute to society through the pursuit of
education, learning and research at the highest international levels of excellence.

www.cambridge.org
Information on this title: www.cambridge.org/9781107018754

First published 2012
First paperback edition 2017

A catalogue record for this publication is available from the British Library

Library of Congress Cataloging-in-Publication data
Di Baldassarre, Giuliano, 1978–
Floods in a changing climate. Inundation modelling / Giuliano Di Baldassarre.
    pages   cm. – (International hydrology series)
Includes bibliographical references and index.
ISBN 978-1-107-01875-4
1. Flood damage prevention.   2. Floodplain management.   3. Floodplains.
4. Hydrogeological modeling.   5. Climatic changes – Environmental
aspects.   I. Title.
TC409.D44   2012
551.48′9011 – dc23   2012015663

ISBN   978-1-107-01875-4   Hardback
ISBN   978-1-108-44675-4   Paperback

Additional resources for this publication at www.cambridge.org/dibaldassarre

To my Family

Water is the cause at times of life or death, or increase of privation, nourishes at times and at others does the contrary; at times has a tang, at times is without savour, sometimes submerging the valleys with great floods. In time and with water, everything changes.

Leonardo da Vinci, *circa* 1500

Contents

Contributing authors

Paul D. Bates (Chapter 8)
School of Geographical Sciences, University of Bristol,
Bristol, United Kingdom

Luigia Brandimarte (Chapter 2)
UNESCO-IHE Institute for Water Education,
Delft, the Netherlands

Timothy J. Fewtrell (Chapter 8)
Willis Research Network, Willis Group,
London, United Kingdom

Jeffrey C. Neal (Chapter 8)
School of Geographical Sciences, University of Bristol,
Bristol, United Kingdom

Ioana Popescu (Chapter 3)
UNESCO-IHE Institute for Water Education,
Delft, the Netherlands

Forewords

Everybody speaks about climate change these days, yet not everybody recognizes that most of the impacts of climate variability will be manifested through, with and by water. Whether one speaks about sea level rise or increasing flood frequencies, or the combined effect of the two in the case of coastal areas, it is water that will be the agent of change, for water connects. It connects environmental systems with the social ones; in fact it connects all the major development objectives as set by the Millennium Development Goals (MDGs) as well as matters related to food and energy security. It is, therefore, critical to understand the response of hydrologic systems to extremes.

How will flooding patterns in general change in response to the global drivers that will have regional, national and even local impacts? Is it indeed only climate variability and change that is the main driver behind changes in flood dynamics? How will flood risks change in relation to the global drivers? And what is indeed the most important driver that will influence flood risk management, say, in forty years when there will be approximately nine billion human beings on the Earth? Will it be climate variability or other global drivers linked to population change, such as land-use changes, migration from rural to urban areas, technology or the expected unprecedented growth of cities? Climate change will likely contribute to increased uncertainty, and thereby risk; however, the main driver that will cause further significant changes in flood dynamics is population increase and the resulting human interventions in the workings of the hydrologic cycle.

The water science community is grappling with a major question: Is it true that the hydrologic cycle is accelerating? Because if this is indeed the case then we have the primary proof that the climate system changes and moves outside the deviations linked to normal climate variability. If that is the case then we have the principal proof that flood frequencies do indeed increase and for that matter the probability of other hydrologic extremes, such as droughts, occurring more often will indeed increase. As a net result we will have more floods.

QED, one would be tempted to say. However, we do not have the solid evidence that the hydrologic cycle indeed accelerates at a global scale. There is no global trend observed yet that would indicate either an increased flow or a decreased flow. There are some rivers where flow patterns display a decreasing trend, while there are others which show an increasing flow tendency. Overall no clear trend can be identified. Even at continental scale the balance between increasing versus decreasing flows seems to be all right.

One reason to come to this conclusion might be that indeed the hydrologic cycle is not accelerating and the overall system is at equilibrium. The other conclusion one might have is somewhat more prosaic: we simply do not know enough about the workings of the hydrologic cycle. One reason behind that is that the statistical hypothesis and tools we use are too weak to detect the change. After all, we are still using a toolbox that contains tools designed to handle stationary processes. That assumption is surely not true any more in our exponentially changing world. The second reason is the age-old issue of data scarcity. First of all, the time series we have are relatively short to make inferences for large time scales that typically characterize climate change, even if the data sets started in the late nineteenth century. Second, the issue of spatial scarcity is even more striking. Take, for instance, the case of Africa where data availability is very scarce due to historical and other reasons. On top of these, there is a third reason why we are facing problems in properly managing flood risk as a function of various drivers, and that is the sizeable gap that exists in our understanding of the relevant processes.

Irrespective of these issues and uncertainties, one thing seems to be quite certain: flood vulnerability and risks will no doubt increase in the coming decades. Owing to the fact that more and more people are moving into flood-prone areas it is no longer sufficient to issue forecasts for the flood hydrographs alone, as the two-dimensional character of flooding will dominate the success of flood management activities.

This fact alone underlines the huge importance of Giuliano Di Baldassarre's present book. What the reader is presented with in this volume is a systematic treatment of flood inundation modelling ranging from the theoretical backgrounds of unsteady flow all the way up to the making and interpretation of floodplain mapping. Di Baldassarre has done very commendable work by putting in one comprehensive framework both the relevant theory and its applications. A great number of examples, ranging from urban flood modelling to the evaluation of floodplain management strategies, and exercises help the understanding of the underlying concepts.

The material presented herein could be used in various teaching courses at different levels and also as a case study book in flood management. Therefore, I would like to recommend this excellent volume wholeheartedly for both academics and practitioners involved in flood management as the knowledge contained in the volume will certainly help reduce the risks of flood inundation and thereby will help in moving towards sustainable water management.

Professor A. Szöllösi-Nagy
Rector
UNESCO-IHE Institute for Water Education

There are scientific issues related to earth sciences that are extremely important for our everyday life and have benefited much from recent research results and improved environmental monitoring. Inundation modelling is an excellent example where the progress is amazing. Scientists have recently been able to deal with the increasing problems related to inundations through an efficient synthesis between technical capabilities, improved computational means and research advances. Indeed, illustrating the above progress, to further help translating research results into technical practice, is an excellent idea and this book does the job in a clear and exhaustive manner.

Water has always been a key driver of social development and therefore living with, and protecting from, water has always been one of the arts of humanity. Today the art is becoming more challenging due to the increasing needs originating from the improving social welfare. The recent flood events that have occurred all over the world have pointed out the urgent necessity to predict how water expands over floodplain and urban areas. Such events clearly show that we are not prepared enough to deal with water flowing over roads and among houses, while recent research results show that such events can be modelled and their effect predicted, by profiting from extraordinarily improved monitoring capabilities. Therefore, the above art, which was recently enriched with important contributions, needs to be supported with new educational tools.

This is the reason why I enthusiastically appreciated the idea of Giuliano Di Baldassarre writing this book. It is uncommon to see a young scientist writing a book, and therefore I am very much delighted to see his signature under this timely and precious contribution. Indeed, it shows that the motivation, preparation and clarity of ideas that support young scientists are an invaluable contribution to science and society. When I read these pages I could not avoid my thoughts pleasingly going back to 6 years ago, remembering the time when Giuliano and I were working together on his Ph.D. research and every day I was impressed by his rigorousness with details and clarity. One of the reasons why research is a very rewarding job is the opportunity to meet extraordinary persons.

What I particularly like in this book is the emphasis that is given to uncertainty estimation for decision-making, which is tackled here with an original approach that makes use of several sources of information. Chapters 5, 6 and 7 bring forward an original contribution that will open the doors to further research activity. In particular, an important issue is highlighted that is often not considered enough, namely, uncertainty in the boundary conditions for hydraulic modelling. Finally, this book emphasizes the opportunity to include social forcing in environmental modelling. Environment and society are linked and conditional on each other: understanding the underlying connections is a fundamental step forward to improving living conditions and, in particular, reducing flood risk.

I warmly address to Giuliano Di Baldassarre my personal appreciation. I also would like to thank all the readers of these few words, which I wrote with great pleasure.

Professor A. Montanari
University of Bologna

Preface

Floodplains are among the most valuable ecosystems for providing goods and services to the environment and supporting biodiversity. At the same time, it is estimated that almost one billion people, the majority of them the world's poorest inhabitants, currently live in floodplains. As a result, flooding is nowadays the most damaging natural hazard worldwide. Damage and fatalities caused by flood disasters are expected to further increase dramatically in many parts of the world because of continuous population growth in floodplains as well as changes in land use and climate.

Over the past decades, I have been looking at different methods – developed by hydrologists, ecologists, engineers and geomorphologists – to observe and analyse floodplain systems. These floodplain models range in complexity from simply intersecting a plane representing the water surface with digital elevation models to sophisticated solutions of the Navier–Stokes equations. Some of these models have been proved to be useful tools in floodplain management, understanding sediment dynamics and flood risk mitigation. For instance, their ability to predict inundation extents can be used to reduce the potential flood damage by supporting more appropriate land use and urban planning, raising the awareness of people living in flood-prone areas, and discouraging new human settlements in floodplains.

Thus, I was really glad when I was contacted by Slobodan Simonovic and given the opportunity to write this book, dealing with floodplain dynamics and inundation modelling, as one of the collection of books within the International Hydrology Series on flood disaster management theory and practice within the context of climate change.

And here we are. This book, *Floods in a Changing Climate: Inundation Modelling* – prepared under the responsibility and coordination of Siegfried Demuth, UNESCO International Hydrological Programme (IHP), Chief of Hydrological Systems and Global Change Section and scientist responsible for the International Flood Initiative (IFI), and Biljana Radojevic, Division of Water Sciences – is intended for graduate students, researchers, members of governmental and non-governmental agencies and professionals involved in flood modelling and management. A number of revision exercises are included in the book to promote more effective learning of concepts within academic environments. Access to online electronic resources including software for one-dimensional (1D) and two-dimensional (2D) hydraulic modelling is also provided.

The book is structured as follows: Introduction, Theory (Part I), Methods (Part II), and Applications (Part III). Throughout the book, particular attention is given to, on the one hand, the challenge of dealing with the estimation of the uncertainty affecting any modelling exercise, and, on the other hand, the opportunity given by the current proliferation of remote sensing data to improve our ability to model floodplain inundation processes. The first part of the book (Chapters 2 and 3) provides a concise, but as comprehensive as possible, mathematical description of the basic hydraulic principles, steady and unsteady flow equations, numerical and analytical solutions. The second part (Chapters 4–7) is the core of the book and its structure reflects the steps necessary for the implementation of hydraulic modelling of floods: data acquisition, model building, model evaluation, and elaboration of model results in a GIS environment. Lastly, the third part (Chapters 8–11) shows four different example applications of flood inundation modelling in a rapidly changing world: analysis of urban floods, changes in flood propagation caused by human activities, changes in stage–discharge rating curves, and evaluation of different floodplain management strategies.

In conclusion, I would like to highlight that this book could not have been made without the kind and substantial contributions of Paul Bates, Luigia Brandimarte, Tim Fewtrell, Jeff Neal, Ioana Popescu, Durga Lal Shrestha, and András Szöllösi-Nagy. Also, I would like to acknowledge my father Domenico Di Baldassarre for kindly drawing some of the figures, as well as Francesco Dottori and Leonardo Alfonso Segura for providing precious support during the book writing process. Lastly, the book includes concepts and thoughts that emerged by interacting with colleagues and friends over the past few years. Here, I feel I must mention, in completely random order: Micah Mukolwe, Alberto Montanari, Philip Tetteh Padi, Doug Alsdorf, Elena Toth, Kun Yan, Stefan Uhlenbrook, Alessio Domeneghetti,

Matt Horritt, Anuar Ali, Elena Ridolfi, Dimitri Solomatine, Semu Moges, Pierluigi Claps, Micha Werner, Jim Freer, Maurizio Mazzoleni, Demetris Koutsoyiannis, Max Pagano, Attilio Castellarin, Mohamed Elshamy, Ann van Griensven, Eman Soliman, Florian Pappenberger, Armando Brath, Patrick Matgen, Keith Beven, Max Kigobe, Salvatore Grimaldi, Preksedis Ndomba, Alessandro Masoero, Nigel Wright, Laura Giustarini, Joseph Mutemi, Huub Savenije, Salvano Briceno, Simone Castiglioni, Slobodan Simonovic, Siegfried Demuth, Yunqing Xuan, Roberto Ranzi, Francesco Laio, Pietro Prestininzi, Matt Wilson, Paolo D'Odorico, Harry Lins, Stefano Barontini, Neil Hunter, Günter Blöschl, and Guy Schumann.

1 Introduction

1.1 FLOODS: NATURAL PROCESSES AND (UN)NATURAL DISASTERS

Since the earliest recorded civilizations, such as those in Mesopotamia and Egypt that developed in the fertile floodplains of the Tigris and Euphrates and Nile rivers, humans have tended to settle in flood-prone areas as they offer favourable conditions for economic development (Di Baldassarre *et al.*, 2010a). However, floodplains are also exposed to flood disasters that might cause severe damage in terms of society, economy, environment and loss of human lives (Figure 1.1).

A flood disaster is said to occur when an extreme event coincides with a vulnerable physical and socio-economic environment, surpassing society's ability to control or survive the consequences. Currently, flood disasters account for half of all deaths caused by natural catastrophes (Ohl and Tapsell, 2000). In 2010, floods were responsible for the loss of more than 8,000 human lives and affected about 180 million people (Figure 1.2; EM-DAT, 2010).

Yet the catastrophic floods that occurred in 2010 (e.g. Pakistan and China) are only the most recent examples of worldwide increasing flood damage. Figure 1.3 shows, for instance, that the number of people affected by floods in the African continent has dramatically increased over the last decades (EM-DAT, 2010). Sadly, similar diagrams can be derived by analysing flood damage and fatalities in other continents.

To mitigate the continuously increasing flood risk the currently proposed approach is integrated flood management (aimed more towards 'living with floods'), which has replaced the more traditional flood defence approach ('fighting floods'). This approach aims to minimize the human, economic and ecological losses from floods while, at the same time, maximizing the social, economic and ecological benefits (UNESCO-IFI, International Flood Initiative). Thus, flood managers should be concerned not only about the reduction of the potential damage of extreme flood events, but also about the protection of floodplains, which are among the most valuable ecosystems for providing goods and services to society and supporting biodiversity (Costanza *et al.*, 1997; Nardi *et al.*, 2006; Opperman *et al.*, 2009).

However, how to implement integrated flood management schemes including the needed capacity development activities in an ever changing world is often unknown and requires research and rethinking of our current approaches (Uhlenbrook *et al.*, 2011). This seems to be true in particular in the developing world, where better flood management is very much needed to limit the societal impacts of floods (Di Baldassarre and Uhlenbrook, 2011).

1.2 DEFINITIONS

Flood is a natural process that can be defined as a body of water which rises to overflow land that is not normally submerged (Ward, 1978). It can be generated by many causes (and combinations thereof) that include: heavy rain, rapid snow/ice melt, glacial lake breaches, ice breakup, debris entrapment, dam breaks, levee breaches, landslide blockages and groundwater rises. The most common types of flood are storm surges, river floods and flash floods. Flood risk is typically defined as the result of the integration of two components, i.e. probability and consequences (Sayers *et al.*, 2002; Simonovic, 2012):

$$Risk = Probability \times Consequences \qquad (1.1)$$

This concept of risk is strictly related to the probability that a flood event of a given magnitude occurs, while consequences are the expected environmental, economic and social losses caused by that flood event. This definition of risk is also used in the recent European Flood Directive 2007/60/EC (European Parliament, 2007) where flood risk is a combination of the probability of a flood event and the potential adverse consequences for human health, the environment, cultural heritage and economic activity.

Another widely used definition of flood risk specifies the two contributions to the consequences caused by a hazardous event,

1

Figure 1.1 River Aniene (Italy) during the March 2011 flooding (photo by Max Pagano).

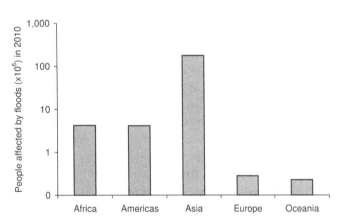

Figure 1.2 Number of people affected by floods in 2010 by continents. Note the use of the logarithmic scale. In order for a disaster to be entered into the OFDA/CRED International Disaster Database at least one of the following criteria has to be fulfilled: (i) 10 or more people reported killed, (ii) 100 people reported affected, (iii) a call for international assistance, (iv) declaration of a state of emergency (EM-DAT, 2010).

i.e. vulnerability and exposure (Sagris *et al.*, 2005; Landis, 2005; UN-ISDR, 2004):

$$Risk = Hazard \times Exposure \times Vulnerability \qquad (1.2)$$

This definition requires that a vulnerable area (from a social, economic or environmental point of view) is actually exposed to

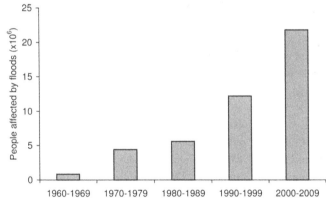

Figure 1.3 Number of people affected by floods in Africa (EM-DAT, 2010).

the hazard. If an event occurs where there is no vulnerability or no exposure, then there is also no risk.

The two definitions of flood risk, (1.1) and (1.2), are clearly interrelated and interchangeable and each of these two definitions has certain advantages in different applications (e.g. Sayers *et al.*, 2002; Landis, 2005; Merz *et al.*, 2007). More details on flood risk management can be found in Simonovic (2012).

As mentioned, facts and hard data clearly indicate that flood risks have increased over the last decades. This dramatic increase may have been caused by a combination of climate and

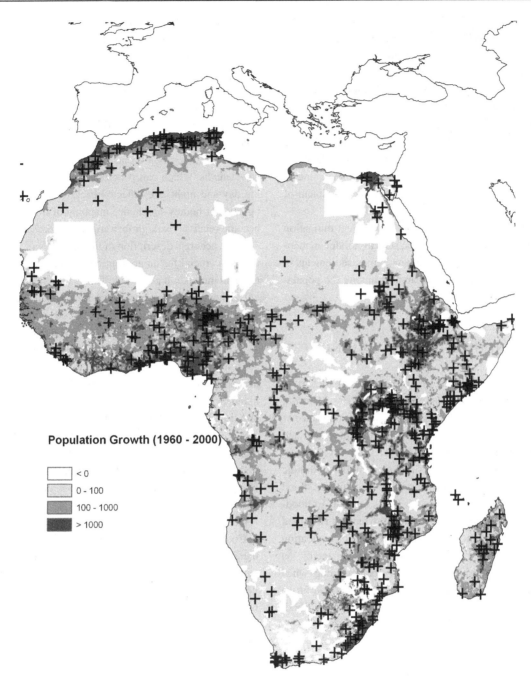

Figure 1.4 Spatial distribution of population growth (as number of inhabitants per cell of 2.5′) and location of the most recent floods (crosses); see also Di Baldassarre *et al.* (2010a).

land-use changes, which may have increased flood probability, and economic and demographic changes, which may have led to increased human vulnerability to extreme hydro-meteorological conditions. For instance, the aforementioned increase of flood losses in Africa was found to be caused by intensive and unplanned urbanization of flood-prone areas, which has played a major role in increasing the potential adverse consequences of floods (Figure 1.4).

In particular, Figure 1.4 shows, at the continental scale, the dynamics of human settlements (i.e. population growth between 1960 and 2000) and the location of the latest floods in Africa (Dartmouth Flood Observatory, 2010) and highlights that most of the recent floods (i.e. period 1985–2009) have occurred where the population has increased more. This is not only the case for the African continent. A dramatic example is the May 2004 flooding of the transboundary River Soliette (Haiti and the Dominican

Republic) where formal and informal human settlements in the floodplain led to dramatically high flood casualties, i.e. more than 1,000 people were killed in addition to many hundreds of people left homeless (Brandimarte *et al.*, 2009).

Lastly, although risk awareness and population dynamics are rather different, intensive urbanization of flood-prone areas is also widely present in more developed countries. For instance, over 12% of the population of the United Kingdom (UK) live on fluvial and coastal floodplains, about half of the population of the Netherlands live below (or close to) mean sea level, and in Hungary about 25% of the population live on the floodplain of the River Danube and its tributaries (BRISK, 2011).

These dramatic figures indicate the need for urgent mitigation actions to tackle the increasing flood risk, such as floodplain mapping, which can help in discouraging new human settlements in flood-prone areas and raising risk awareness among the population living in floodplains (Padi *et al.*, 2011).

1.3 FLOOD INUNDATION MODELLING

Flood inundation models are numerical tools able to simulate river hydraulics and floodplain inundation processes (Horritt *et al.*, 2007). In recent years, the increased socio-economic relevance of river flood studies and a shift of these studies towards integrated flood risk management concepts have triggered the development of various methodologies for the simulation of the hydraulic behaviour of river systems (see Chapter 5). In particular, flood inundation models have been proved to be useful tools in floodplain management, understanding sediment dynamics and flood risk mitigation. For instance, their ability to predict inundation extents can be used to reduce the potential flood damage by: (i) supporting a more appropriate land use and urban planning (when present); (ii) raising the awareness of people living in flood-prone areas; and (iii) discouraging new human settlements in floodplains.

Flood modellers are well aware that a significant approximation affects the output of their models. Uncertainty is caused by many sources of error that propagate through the model and therefore affect its output. Three main sources of uncertainty have been identified (Götzinger and Bardossy, 2008): (i) observation uncertainty, which is the approximation in the observed hydrologic variables used as input or calibration data (e.g. rainfall, temperature and river discharge); (ii) parameter uncertainty, which is induced by imperfect model calibration; (iii) model structural uncertainty, which originates in the inability of models to perfectly schematize the physical processes involved. In recent years, there has been an increasing interest in assessing uncertainty in flood inundation modelling, analysing its possible effects on floodplain mapping, and making a more efficient use of data to constrain uncertainty (Di Baldassarre *et al.*, 2009a).

Nowadays, a great opportunity to reduce the uncertainty of models is offered by the increasing availability of distributed remote sensing data, which has led to a sudden shift from a data-sparse to a data-rich environment for flood inundation modelling (Bates, 2004a). For instance, flood extent maps derived from remote sensing are essential calibration data to evaluate inundation models (Horritt *et al.*, 2007). From space, satellites carrying synthetic aperture radar (SAR) sensors are particularly useful for monitoring large flood events (Aplin *et al.*, 1999). In fact, radar wavelengths, which can penetrate clouds and acquire data during day and night, are reflected back to the antenna by smooth open water bodies, and hence mapping of flood extent areas has become relatively straightforward (Di Baldassarre *et al.*, 2011a). Also, an accurate description of the geometry of rivers and floodplains is crucial for an appropriate simulation of flood propagation and inundation processes. This is currently allowed by modern techniques for topographical survey, such as airborne laser altimetry (LiDAR; e.g. Cobby *et al.*, 2001), that enable numerical descriptions of the morphology of riverbanks and floodplain areas with planimetric resolution of 1 m and finer. The elevation accuracy of these LiDAR data is between 5 and 15 cm, which makes this type of topographic data suitable to support flood inundation modelling. Lastly, it is worth mentioning that, in the last decade, there has been dissemination of topographic data that are freely and globally available, such as the space-borne digital elevation model (DEM) derived from the Shuttle Radar Topography Mission (SRTM), which has a geometric resolution of 3 arc seconds (LeFavour and Alsdorf, 2005) and covers most of the land surfaces that lie between 60° N and 54° S latitude (Figure 1.5).

Confirmation of the utility of globally and freely available data therefore indicates the potential to remove an important obstacle currently preventing the routine application of models to predict flood hazards globally, and potentially allows such technology to be extended to developing countries that have not previously been able to benefit from flood predictions. However, clear guidelines to fully and properly utilize the current 'flood of data' (Lincoln, 2007) are still to be developed (Di Baldassarre and Uhlenbrook, 2011).

1.4 CLIMATE AND FLOODS

There is global concern that flood losses might grow further in the near future because of many factors, such as changing demographics, technological and socio-economic conditions, industrial development, urban expansion and infrastructure construction, unplanned human settlement in flood-prone areas, climate variability and change (full report of the Scientific and Technical Committee, UN-ISDR, 2009).

Figure 1.5 This world map shows the SRTM-derived flow accumulation area (greyscale; from black to white), i.e. the amount of basin area draining into each cell. Larger rivers are recognizable as white areas.

In recent years, a large part of the scientific community has made efforts in analysing the impact of climate change on water resources and proposing adaptation strategies (Wilby *et al.*, 2008). The usual framework of this type of studies can be summarized as follows (Di Baldassarre *et al.*, 2011b): (i) choice of one or more scenarios of the IPCC (Intergovernmental Panel on Climate Change) special report on emission scenarios (Bates *et al.*, 2008), which depend on the future economy and energy use policies; (ii) choice of one or more global climate models (GCM); (iii) downscaling of the GCM output to the specific river basin scale; (iv) use of the downscaled GCM outputs as inputs for a hydrologic model; and (v) analysis of hydrologic model results by comparing them to the corresponding results related to the current climate or different possible future climates (see also Mujumdar and Kumar, 2012; Teegavarapu, 2012). This approach has become very popular as it potentially allows the quantification of changes in floods, flow duration curves, and the appropriate part of the hydrologic cycle. However, it should be noted that different techniques may lead to opposing trends and contradicting recommendations for policy-makers (Blöschl and Montanari, 2010).

It has been customary for water communities to use climate model outputs as quantitative information for assessing climate impacts on water resources and, in particular, flood risk management (Simonovic, 2012). However, caution is always needed in considering certain modelling aspects, such as: (i) the choice of the particular model or set of global models to use; (ii) domain configurations for regional models; (iii) choosing appropriate model physics especially for those handling moist convective processes related to reproducing observational climatology and inter-annual features of regional and local precipitation. Di

Baldassarre *et al.* (2011b) indicated the need for good practice in climate impact studies. This practice should include the following requirements: (i) results should not be presented in a simplified way assuming a one-way cause–effect relationship; (ii) ensembles of several climate model projections should be used to reflect their large variability; (iii) the performance of the models applied to historical data should be provided; (iv) appropriate downscaling techniques should be used and the underlying assumptions should be reported; and (v) appropriate uncertainty analysis techniques should be applied to the entire modelling chain. Blöschl and Montanari (2010) recommended that impact studies should not only present the assumptions, results and interpretation, but also provide a clear explanation of 'why' certain changes are projected by the applied models. The idea is that we should not trust that the results are valid unless we understand why an impact study projects changes in a given hydrologic variable.

More details of climate impact on floods are reported in the other volumes of the book series on Floods in a Changing Climate (Mujumdar and Kumar, 2012; Simonovic, 2012; Teegavarapu, 2012). As far as this book is concerned, the two main changes that flood inundation modellers should consider are the changes in the frequency (and magnitude) of floods and sea level rise, which impact the boundary condition of flood inundation models (see Chapters 4–7).

For what concerns changes in the frequency of floods, Wilby *et al.* (2008) recently recommended precautionary allowances for the design flood (i.e. peak river flow corresponding to a given return period; Chapter 7) of +10% in 2025, and +20% in 2085 (Wilby *et al.*, 2008). It should be noted that, given the aforementioned uncertainty related to climate impact on floods, these

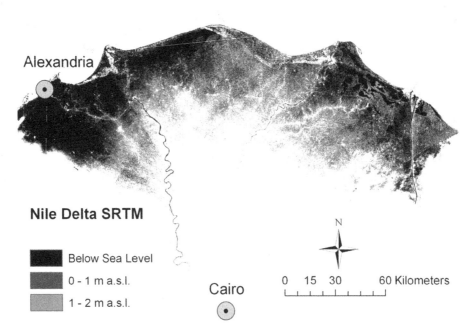

Figure 1.6 Potential impact of sea level rise on the Nile Delta, Egypt. The SRTM topography is classified to show the regions that are currently below sea level (black areas), and the territories that might be potentially flooded by sea level rise (greyscale; see legend).

adjustment factors might be updated in the near future as a result of currently ongoing research.

For sea level rise, Bates *et al.* (2008) indicated, 'The average rate of sea-level rise during the 21st century is very likely to exceed the 1961–2003 average rate (around 1.8 mm/yr)' and is expected to be characterized by 'substantial geographical variability'. More recently, Church and White (2011) pointed out, 'Since the start of the altimeter record in 1993, global average sea level rose at a rate near the upper end of the sea level projections of the Intergovernmental Panel on Climate Change's Third and Fourth Assessment Reports.'

To illustrate the potential impact of sea level rise, Figure 1.6 shows, as an example, the Nile Delta (Egypt, Mediterranean Sea). The Nile Delta covers only 2% of Egypt's territory, but is home to 41% of the Egyptian population and includes 63% of its agricultural land (Hereher, 2010). Figure 1.6 was derived by following a simplified approach, based on the use of SRTM topography (see Chapter 4) with the only objective being to illustrate the potential increase of exposure to coastal flooding related to sea level rise.

UNEP/GRID (2000) and Hereher (2010), which investigated the vulnerability of the Nile Delta with more elaborate techniques, pointed out that sea level rise might seriously affect the Nile Delta as it would lead to shoreline erosion, contamination of lagoons, deterioration of water quality, and inundation of much valuable and productive agricultural land (Figure 1.6). Sadly, similar issues are being experienced in many coastal regions and deltas of the world (UNEP/GRID, 2000).

1.5 PROBLEMS ADDRESSED BY THIS BOOK

This book deals with numerical models able to simulate flood propagation and inundation processes. It provides a dissertation about the state-of-the-art in hydraulic modelling of floods as part of the flood risk management exercise (Simonovic, 2012).

More specifically, the first part of the book (Chapters 2 and 3) provides a concise description of the basic hydraulic principles, steady and unsteady flow equations, and their numerical and analytical solutions. Chapter 4 discusses different data sources to support flood inundation modelling by describing traditional ground-surveyed data (e.g. cross sections, hydrometric data) as well as remotely sensed data (e.g. satellite and airborne images). Chapter 5 deals with model implementation in both theoretical and practical terms. In particular, the chapter introduces the principle of parsimony and the main criteria behind the selection of the most appropriate hydraulic model for simulating flood inundation. Then, numerical tools for flood inundation modelling are classified and briefly described. The chapter also deals with the most common issues related to model building, such as the schematization of model geometry and the parameterization of flow resistance. Chapter 6 discusses the evaluation of flood inundation models. After the introduction of basic concepts, the chapter presents performance measures that are commonly used to compare model results and observations. The calibration and validation of hydraulic models is also discussed.

Lastly, the chapter introduces methodologies recently proposed in the scientific literature that can be used to cope with uncertainty in hydraulic modelling. Chapter 7 deals with the use of model results in GIS environments and describes the necessary steps to build flood hazard maps. It also includes a comparison of deterministic and probabilistic approaches to mapping floodplain areas.

The last part of the book (Chapters 8–11) reports four different example applications. In these examples, flood inundation models are used to: simulate urban flooding, evaluate changes on flood propagation caused by human activities, estimate changes of the stage–discharge rating curve, and compare different floodplain management strategies.

The overall aim of this book is to support an efficient and appropriate implementation of flood inundation models, which have been proved to be useful and essential tools for flood management under climate change. However, it should be noted that modelling flood propagation and inundation processes is only a small part of the risk management exercise. In this context, Szöllösi-Nagy (2009) stated, 'Models play the same role as the heart in human body. Small, but one just cannot exist without it.' It is also worth quoting here the message of Kofi Annan to the World Water Day (2004):

Modern society has distinct advantages over those civilizations of the past that suffered or even collapsed for reasons linked to water. We have great knowledge, and the capacity to disperse that knowledge to the remotest places on earth. We are also beneficiaries of scientific leaps that have improved weather forecasting, agricultural practices, natural resources management, disaster prevention, preparedness and management... But only a rational and informed political, social and cultural response – and public participation in all stages of the disaster management cycle – can reduce disaster vulnerability, and ensure that hazards do not turn into unmanageable disasters.

Part I
Theory

2 Theoretical background: steady flow

Contributing author: Luigia Brandimarte

This chapter provides a concise description of the basic hydraulic principles, such as Bernoulli's principle, with a focus on steady flow equations and backwater computations for open channel hydraulics. Revision exercises complete the chapter.

2.1 UNIFORM FLOW

Uniform flows in open channels are characterized by constant water depth (h) and constant mean velocity (V) along the flow direction (s):

$$\frac{\partial h}{\partial s} = 0$$
$$\frac{\partial V}{\partial s} = 0 \qquad (2.1)$$

Uniform flow can occur only in cylindrical river beds, with constant bed slope and constant discharge. In a channel with a given cross section and given roughness characteristics of the cross section, a given discharge will flow in uniform flow conditions in that given cross section only with a certain mean velocity. This is the mean velocity at which the friction slope and the bed slope are parallel.

In uniform flow conditions, the relation between the mean velocity, V_0, and the characteristics of the flow and the river reach can be expressed by the uniform flow equation, usually given by either the Chezy or the Manning formula:

$$V_0 = C_0 \sqrt{R_0 S_0} \qquad (2.2)$$

$$V_0 = \frac{1}{n} R_0^{\frac{2}{3}} S_0^{\frac{1}{2}} \qquad (2.3)$$

where S_0 is the bed slope (replacing the friction slope, as in uniform flow conditions $S_0 = S_f$), R_0 is the hydraulic radius corresponding to the h_0 water depth, C_0 is the Chezy coefficient function of the roughness and the hydraulic radius, n is Manning's roughness coefficient.

The water depth corresponding to a uniform flow is called the normal depth or uniform flow depth, h_0. For a given discharge and channel geometry (cross section, A_0, and bed slope), the Chezy or

Table 2.1 *The terms of Bernoulli's principle*

Head	Definition	Associated with
Potential head	z	Gravitational potential energy
Pressure head	p/γ	Flow work
Velocity head	$V^2/2g$	Kinetic energy

Manning formula can be used to compute the normal depth. By using Manning's equation, in terms of discharge Q:

$$Q = Q(h_0) = A_0 V_0 = \frac{1}{n} A_0 R_0^{\frac{2}{3}} S_0^{\frac{1}{2}} \qquad (2.4)$$

2.2 SUBCRITICAL AND SUPERCRITICAL FLOWS

Bernoulli's principle (see Table 2.1) states that for a steady, incompressible, perfect flow the total head, H, along a streamline, S, is constant. The total head, H, is the sum of the potential head, pressure head and velocity head (Figure 2.1):

$$H(S) = z + \frac{p}{\gamma} + \frac{V^2}{2g} = \text{Constant} \qquad (2.5)$$

$$H(S) = z_A + \frac{p_A}{\gamma} + \frac{V_A^2}{2g} = z_B + \frac{p_B}{\gamma} + \frac{V_B^2}{2g}$$
$$= z_C + \frac{p_C}{\gamma} + \frac{V_C^2}{2g} = \text{Constant} \qquad (2.6)$$

Bernoulli's principle shows the possibilities and ways of transforming the mechanical energy of a liquid from one form to another (an increase of the velocity due to a decrease of the elevation). Thus, if the pressure distribution is hydrostatic, the total head, H, with respect to the datum (Figure 2.2), at a given section I of an open channel having bed slope α, may be written as

$$H = z_P + h_P \cos \alpha + \frac{V_P^2}{2g} \qquad (2.7)$$

where z is the elevation of point P above the given datum, h is the depth of point P below the water surface measured along the

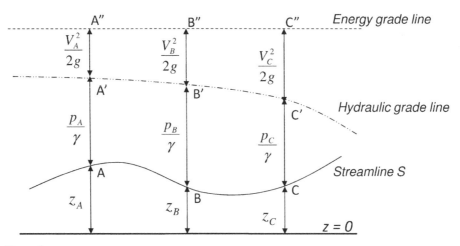

Figure 2.1 The Bernoulli equation.

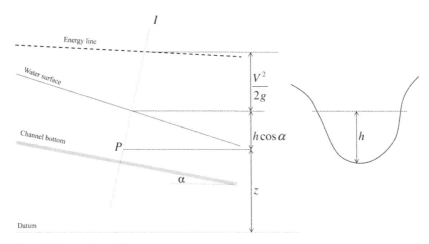

Figure 2.2 Energy in gradually varied open channel flow.

channel bottom, $V^2/2g$ is the velocity head of the flow in the streamline passing through P.

A uniform distribution of the velocity for each streamline passing through the cross section can be attained in the ideal case of uniform flow, with the streamlines parallel to the channel bottom: the non-uniform distribution of velocity in a given channel section in gradually varied flows is taken into account by using an energy coefficient in the velocity head.

Under this assumption, the total energy in cross section I is

$$H = z + h \cos \alpha + \alpha \frac{V^2}{2g} \qquad (2.8)$$

which, for a channel of small slope $\alpha \cong 0$, becomes

$$H = z + h + \alpha \frac{V^2}{2g} \qquad (2.9)$$

The specific energy, E, is the energy at a channel section measured with respect to the channel bottom (not to the datum). Thus, $z \cong 0$

in equation (2.9) and the specific energy is the sum of the water depth and the velocity head:

$$E = h + \alpha \frac{V^2}{2g} = h + \frac{\alpha Q^2}{2g A(h)^2} \qquad (2.10)$$

For a given geometry of the cross section, $A(h)$, and for a given constant discharge, Q, the specific energy, $E(h)$, is a function of the water depth, h, only.

The specific energy curve plots the specific energy, E, against the water depth, h: if the water depth decreases, approaching zero, then the velocity head increases, approaching the vertical axis asymptotically; if the water depth in the section increases, then the velocity head decreases and approaches zero, with the specific energy increasing as the water depth increases, approaching the 45 degree asymptote (if the bottom slope is small).

The $E(h)$ function has a minimum for a certain value of the water depth, at which the first descending limb changes its derivative.

The minimum of the $E(h)$ function can be computed by differentiating the function with respect to h:

$$\frac{dE}{dh} = 1 - \frac{\alpha Q^2}{g A(h)^3} \frac{dA}{dh} = 0 \qquad (2.11)$$

The differential water area near the free surface, dA, is given to the variation of the water depth, dh, by the surface width, B (function of h), $dA = B\,dh$, thus,

$$\frac{dE}{dh} = 1 - \frac{\alpha Q^2}{g A(h)^3} B(h) = 0 \qquad (2.12)$$

Thus the minimum of the $E(h)$ function is given by the value of water depth, h, at which

$$\frac{A(h)^3}{B(h)} = \frac{\alpha Q^2}{g} \qquad (2.13)$$

This value of water depth is known as the critical depth, k, and the critical state of flow is that state at which the specific energy is a minimum for a given discharge; the critical velocity is the mean velocity at the critical state of flow. The critical velocity can be expressed (2.13) by

$$V_k = \frac{Q}{A(k)} = \sqrt{g\frac{A(k)}{\alpha B(k)}} \qquad (2.14)$$

In the case of a rectangular cross section, the computation of the critical depth can be easily made by replacing in equation (2.13) $A = Bh$ and by referring to the discharge per unit width, $q = Q/B$:

$$k = \sqrt[3]{\frac{\alpha Q^2}{g B^2}} = \sqrt[3]{\frac{\alpha q^2}{g}} \qquad (2.15)$$

And from equation (2.14), given that for a rectangular section the ratio A/B that represents the average water depth, h_m, is $A/B = k_m = (Bk)/B = k$, the critical velocity is

$$V_k = \sqrt{g\frac{k}{\alpha}} \qquad (2.16)$$

Thus, from equation (2.10) the minimum value of the specific energy for a given discharge in a rectangular section is given by

$$E_{min} = E_k = \left(h + \frac{\alpha V^2}{2g}\right) = \left(k + \frac{k}{2}\right) = \frac{3}{2}k \quad (2.17)$$

Figure 2.3 is the plot of the specific energy versus the water depth for a given discharge: this means that each point of the curve represents a state of the flow for that given discharge. The minimum of the curve represents the critical state flow, which splits the curve into two characteristic states of flow:

- supercritical flow: water depth less than the critical depth $(h < k)$ and thus mean velocity greater than the critical velocity $(V > V_k)$;
- subcritical flow: water depth greater than the critical depth $(h > k)$ and thus mean velocity less than the critical velocity $(V < V_k)$.

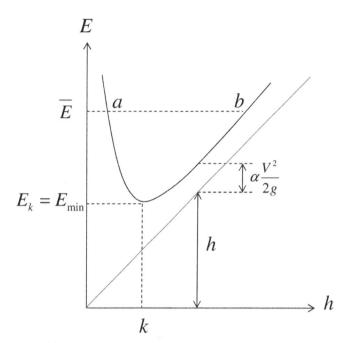

Figure 2.3 Specific energy curve.

Furthermore, it can be noticed on Figure 2.3 that for a specific energy $\bar{E}$ the given discharge Q can flow in supercritical or subcritical conditions (points a and b). As the $\bar{E}$ level decreases, the two points a and b will get closer until they merge into the critical point, k, when reaching the state of minimum specific energy.

In open channels, the transition from subcritical to supercritical flow occurs in a natural, low-loss and smooth way, with the establishment of the critical flow. This is the case, for example, for the sudden change from mild to steep slope (Figure 2.4).

On the other hand, the transition from supercritical to subcritical flow is a highly turbulent phenomenon, which occurs with a strong dissipation of energy. The transition region, where the flow varies rapidly, is characterized by large-scale turbulence, vortices, surface waves and energy dissipation and is called a hydraulic jump.

To analyse the hydraulic jump phenomenon, the momentum equation is used: the change in momentum flux across the control volume equals the sum of the forces acting on the control volume. Let us apply the momentum equation to the control volume 1–2 in the flow direction (Figure 2.5). We can neglect the component of the weight force, W, in the flow direction and the friction force, Fr. Because of the stationarity of the phenomenon, the local inertia force is negligible and the only forces to be considered are the pressure forces on the upstream, Fp_1, and downstream, Fp_2, control sections and the momentum flux across the two sections, Fm_1 and Fm_2, in the unit time. Thus,

$$Fp_1 + Fm_1 = Fp_2 + Fm_2 \qquad (2.18)$$

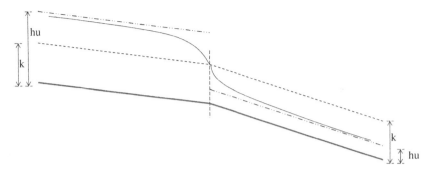

Figure 2.4 Transition from subcritical to supercritical flow condition.

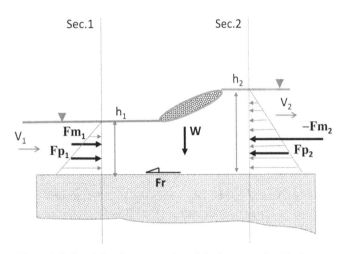

Figure 2.5 Sketch for the computation of the forces involved in the momentum equation.

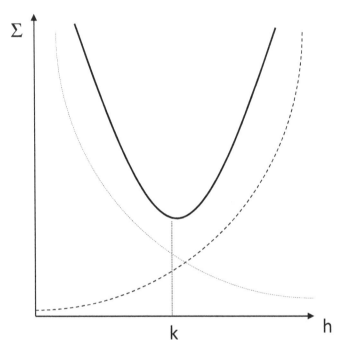

Figure 2.7 Specific forces curve.

and the sum Σ of the two terms is a function of h, when the discharge is constant:

$$\Sigma = Fp + Fm = \gamma Ah_G + \rho \frac{Q^2}{A} \qquad (2.20)$$

From equation (2.20) it is easy to analyse the behaviour of the two terms, Fp and Fm, when h varies. For h null, Fp goes to zero, and with increasing h, Fp increases up to infinity. The momentum flux, Fm, goes as the inverse of the area of the wet section; thus, Fm goes to zero when h goes to infinity and goes to infinity as h goes to zero. The sum of the two terms, Σ, approaches asymptotically to infinity when h goes both to zero and to infinity (Figure 2.7).

The minimum value of the function $\Sigma = f(h)$ can be obtained by deriving the function with respect to h and searching for the value of h that gives the derivative equal to zero.

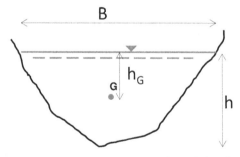

Figure 2.6 Generic cross section and centre of mass below free water surface.

For a generic river cross section having wetted area A, width of the free surface B, water depth h and h_G the depth of the centre of mass (barycentre) below the free surface (Figure 2.6), the pressure force and the momentum flux can be written as

$$Fp = \gamma Ah_G$$
$$Fm = \rho QV = \rho \frac{Q^2}{A} \qquad (2.19)$$

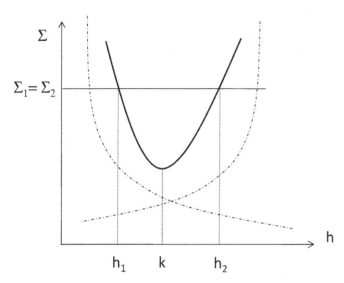

Figure 2.8 Conjugate depths at hydraulic jump location.

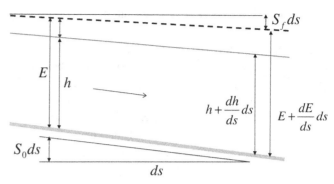

Figure 2.9 Sketch for deriving the gradually varied flow equation.

2.3 WATER SURFACE PROFILES

Let us analyse a flow under the following assumptions: non-uniform flow, steady flow, small bed slope, gradually varied flow; and let us consider an elementary length ds of an open channel, where the discharge Q can be considered constant (no flow incomes or outcomes) and the flow resistance at a given section is that of a uniform flow with the same depth and discharge. The bed slope, S_0, defined as the sinus of the slope angle θ, is assumed to be positive if it descends in the s direction (flow direction) and to be negative if it ascends in the s direction. If we assume, as in Figure 2.9, that the bed slope is positive, then in the elementary length ds, the bed slope drops down by $S_0 \, ds$; the total head line drops down by $S_f \, ds$, where S_f is the friction slope, the head loss per unit length due to the friction along ds. The water surface profile, which represents the piezometric line, can either be ascendant or descendent in the flow direction.

In the simple case of a rectangular channel, equation (2.20) can be written as

$$\Sigma = \gamma A h_G + \rho \frac{Q^2}{A} = \frac{1}{2}\gamma B h^2 + \rho \frac{Q^2}{Bh} \qquad (2.21)$$

Taking the derivative with respect to h, one obtains

$$\frac{d\Sigma}{dh} = \gamma Bh - \rho \frac{Q^2}{Bh^2} \qquad (2.22)$$

The value of h that sets equation (2.22) equal to zero is

$$\frac{d\Sigma}{dh} = 0 \Rightarrow \gamma Bh - \rho \frac{Q^2}{Bh^2} \Rightarrow h = k = \sqrt[3]{\frac{Q^2}{gB^2}} \qquad (2.23)$$

Thus, the minimum of the $\Sigma = f(h)$ function is at the critical depth. The critical depth point divides Figure 2.7 into two sectors: the $h < k$ reach, for the supercritical flow regimes and the $h > k$ reach, for the subcritical flow regimes.

According to equation (2.18), the two depths h_1 and h_2 corresponding to the supercritical flow in section 1 and subcritical flow in section 2 of the control volume (Figure 2.5), have the same value of the Σ ($= Fp + Fm$) terms (Figure 2.8). They are usually referred to as conjugate depths.

Applying equation (2.20) to a rectangular section in terms of unit width, $q = Q/B$, one obtains

$$\Sigma_1 = \frac{1}{2}\gamma h_1^2 + \rho \frac{q^2}{h_1} = \Sigma_2 = \frac{1}{2}\gamma h_2^2 + \rho \frac{q^2}{h_2} \qquad (2.24)$$

Equation (2.24) provides a useful relationship between the conjugate depths in terms of the Froude number at the upstream (supercritical) section:

$$\frac{h_2}{h_1} = \frac{1}{2}\left(-1 + \sqrt{1 + 8Fr_1^2}\right) \qquad (2.25)$$

Equation (2.25), for the specific case of a rectangular section, allows one to compute any of the two conjugate depths at the hydraulic jump, once the other one is known.

Following the scheme in Figure 2.9, the energy equation, in terms of mean specific energy, E, along a streamline in the s direction, can be written as

$$S_0 ds + E = E + \frac{dE}{ds}ds + S_f ds \qquad (2.26)$$

and thus,

$$\frac{dE}{ds} = S_0 - S_f \qquad (2.27)$$

Equation (2.27) shows us that along the flow direction the specific energy, E, increases if the elevation of the bottom channel decreases and decreases because of the friction.

For prismatic channels, the specific energy, E, is a function of s through the flow depth h only, thus equation (2.27) can be written as

$$\frac{dE}{ds} = \frac{dE}{dh}\frac{dh}{ds} = S_0 - S_f \qquad (2.28)$$

From the above equation (2.28) one can derive the equation that gives the slope of the water surface with respect to the channel bottom:

$$\frac{dh}{ds} = \frac{S_0 - S_f}{\frac{dE}{dh}} \qquad (2.29)$$

If $\frac{dh}{ds} = 0$ there is no change of the flow depth along the s direction and the slope of the water surface is equal to slope of the bottom. If $\frac{dh}{ds} > 0$ the water surface is rising along the s direction and the slope of the water surface is less than the slope of the bottom. If $\frac{dh}{ds} < 0$ the water surface is lowering along the s direction and the slope of the water surface is greater than the slope of the bottom. Thus, in order to derive the water surface profile along the flow direction, it is useful to analyse separately the numerator, N, and the denominator, D, of the second member of equation (2.29), which will give us the sign of the $\frac{dh}{ds}$ derivative.

The numerator $N = S_0 - S_f$ is equal to zero for uniform flow conditions, where the friction slope is parallel to the bed slope ($S_0 = S_f$) and $\frac{dh}{ds} = 0$.

Given the assumption that the flow resistance at a given section is that of a uniform flow with the same depth and discharge, we can for instance use the Chezy formula to express the friction slope as a function of flow depth:

$$S_f = \frac{V^2}{C^2 R} = \frac{Q^2}{C^2 R A^2} \tag{2.30}$$

Since the flow depth, h, is in the terms in the denominator, S_f increases when h decreases and decreases when h increases.

Thus:

$$N = S_0 - S_f = 0 \quad \text{if } h = h_{\text{uniform flow}} \tag{2.31}$$

$$N = S_0 - S_f > 0 \quad \text{if } h > h_{\text{uniform flow}} \tag{2.32}$$

$$N = S_0 - S_f < 0 \quad \text{if } h < h_{\text{uniform flow}} \tag{2.33}$$

The denominator of equation (2.29) shows the variation of the specific energy with the flow depth for a given discharge:

$$D = \frac{dE}{dh} = 0 \quad \text{for } h = k \text{ (critical flow condition)} \tag{2.34}$$

$$D = \frac{dE}{dh} > 0 \quad \text{for } h > k \text{ (subcritical flow condition)} \tag{2.35}$$

$$D = \frac{dE}{dh} < 0 \quad \text{for } h < k \text{ (supercritical flow condition)} \tag{2.36}$$

Let us discuss the water surface profiles in mild (M1, M2 and M3) and steep slopes (S1, S2 and S3) separately.

Mild slope channel, $S_0 < S_k$
M1 profile: profile behind a reservoir in natural open channels

This profile occurs in the area above the normal depth line (Figure 2.10; see also Table 2.2): $h > h_u > k$. If we analyse the sign of equation (2.29), we can observe that

$$N = S_0 - S_f > 0 \tag{2.37}$$

because $h > h_u$, and

$$D = \frac{dE}{dh} > 0 \tag{2.38}$$

because $h > k$ (subcritical flow condition).

Table 2.2 *Water surface profiles for gradually varied flows: mild slope channels*

$h > h_u > k$	$N = S_0 - S_f > 0$ $D = \frac{dE}{dh} > 0$	$\frac{dh}{ds} > 0$	M1 (Figure 2.10)
$h_u > h > k$	$N = S_0 - S_f < 0$ $D = \frac{dE}{dh} > 0$	$\frac{dh}{ds} < 0$	M2 (Figure 2.11)
$h_u > k > h$	$N = S_0 - S_f < 0$ $D = \frac{dE}{dh} < 0$	$\frac{dh}{ds} > 0$	M3 (Figure 2.12)

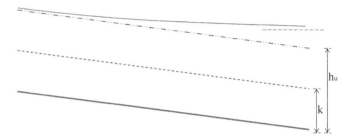

Figure 2.10 Mild slope channel: M1 profile.

Thus, $\frac{dh}{ds} > 0$ and the water surface profile rises in the direction of the flow. Going upstream, the flow depth decreases and the upstream end of the profile is tangent to the normal depth line since $\frac{dh}{ds} = 0$ for $h = h_u$. Going downstream, the flow is tangent to the horizontal since the flow depth increases (theoretically, up to infinity), the flow resistance decreases and the numerator N tends to S_0; the denominator D tends to 1 because h tends to ∞, thus $\frac{dh}{ds}$ tends to S_0.

M2 profile: profile upstream of a sudden expansion of the cross section

This profile occurs in the area between the normal depth line and the critical depth line (Figure 2.11): $h_u > h > k$. If we analyse the sign of equation (2.29), we can observe that

$$N = S_0 - S_f < 0 \tag{2.39}$$

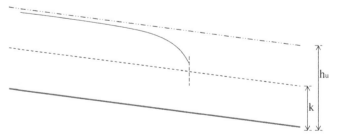

Figure 2.11 Mild slope channel: M2 profile.

because $h < h_u$ and

$$D = \frac{dE}{dh} > 0 \qquad (2.40)$$

because $h > k$ (subcritical flow condition).

Thus, $\frac{dh}{ds} < 0$ and the water surface lowers along the s direction. Going upstream the flow depth increases and becomes tangent to the normal depth line since $\frac{dh}{ds} = 0$ when $h = h_u$. Going downstream, the water surface profile decreases and ends tangent to a vertical line when the flow depth becomes equal to the critical depth, since for $h = k$ the denominator

$$D = \frac{dE}{dh} = 0 \quad \text{and thus} \quad \frac{dh}{ds} = \infty \qquad (2.41)$$

M3 profile: profile below a sluice gate

This profile occurs in the area below the critical depth line (Figure 2.12): $h_u > k > h$. If we analyse the sign of equation (2.29), we can observe that

$$N = S_0 - S_f < 0 \qquad (2.42)$$

because $h < h_u$ and

$$D = \frac{dE}{dh} < 0 \qquad (2.43)$$

because $h < k$ (supercritical flow condition).

Thus $\frac{dh}{ds} > 0$, and the water surface rises along the s direction. Going downstream, the flow depth increases and becomes tangent to a vertical line when the flow depth becomes equal to the critical depth, since, for $h = k$, $D = \frac{dE}{dh} = 0$ and thus $\frac{dh}{ds} = \infty$. Going upstream, h decreases and the theoretical upstream profile will

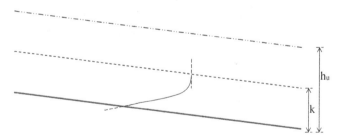

Figure 2.12 Mild slope channel: M3 profile.

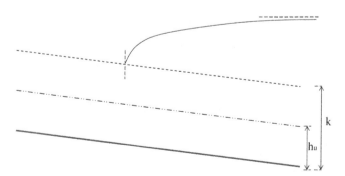

Figure 2.13 Steep slope channel: S1 profile.

Table 2.3 *Water surface profiles for gradually varied flows: steep slope channels*

	$N = S_0 - S_f > 0$		
$h > k > h_u$	$D = \dfrac{dE}{dh} > 0$	$\dfrac{dh}{ds} > 0$	S1 (Figure 2.13)
	$N = S_0 - S_f > 0$		
$k > h > h_u$	$D = \dfrac{dE}{dh} < 0$	$\dfrac{dh}{ds} < 0$	S2 (Figure 2.14)
	$N = S_0 - S_f < 0$		
$k > h_u > h$	$D = \dfrac{dE}{dh} < 0$	$\dfrac{dh}{ds} > 0$	S3 (Figure 2.15)

intersect the bottom of the channel and h values would become negative, with no physical meaning.

Steep slope channel, $S_0 > S_k$

S1 profile: profile behind a reservoir in steep channels

This profile occurs in the area above the critical depth line (Figure 2.13; see also Table 2.3): $h > k > h_u$, which is the only possible subcritical flow in steep slope channels. If we analyse the sign of equation (2.29), we can observe that

$$N = S_0 - S_f > 0 \qquad (2.44)$$

because $h > h_u$ and

$$D = \frac{dE}{dh} > 0 \qquad (2.45)$$

because $h > k$ (subcritical flow condition).

Thus, $\frac{dh}{ds} > 0$ and the water surface profile rises in the direction of the flow. Going upstream, the water surface profile decreases and becomes tangent to a vertical line when the flow depth becomes equal to the critical depth, since, for $h = k$, $D = \frac{dE}{dh} = 0$ and thus $\frac{dh}{ds} = \infty$. Going downstream, the flow is tangent to the horizontal since the flow depth increases (theoretically, up to infinity), the flow resistance decreases and the numerator N tends to S_0; the denominator D tends to 1 because h tends to ∞, thus $\frac{dh}{ds}$ tends to S_0.

S2 profile: profile on the steep slope side of a channel that changes from mild to steep slope

This profile occurs in the area between the critical depth line and the normal depth line (Figure 2.14): $k > h > h_u$, which is a supercritical flow. If we analyse the sign of equation (2.29), we can observe that

$$N = S_0 - S_f > 0 \qquad (2.46)$$

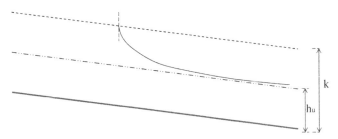

Figure 2.14 Steep slope channel: S2 profile.

because $h > h_u$ and

$$D = \frac{dE}{dh} < 0 \qquad (2.47)$$

because $h < k$ (supercritical flow condition).

Thus, $\frac{dh}{ds} < 0$ and the water surface lowers along the s direction. At the upstream end, the flow depth has a vertical slope at the critical depth; going downstream the water surface profile lowers and becomes tangent to the normal depth line.

S3 profile: profile below a sluice gate with the depth of the outflowing water less than the normal depth on a steep slope

This profile occurs in the area below the uniform depth line (Figure 2.15): $k > h_u > h$, which is a supercritical flow. If we analyse the sign of equation (2.29), we can observe that

$$N = S_0 - S_f < 0 \qquad (2.48)$$

because $h < h_u$ and

$$D = \frac{dE}{dh} < 0 \qquad (2.49)$$

because $h < k$ (supercritical flow condition).

Thus $\frac{dh}{ds} > 0$, and the water surface rises along the s direction. Going downstream, the flow depth increases and becomes tangent to the normal depth line; going upstream, h decreases and the theoretical upstream profile will intersect the bottom of the channel and h values would become negative, with no physical meaning.

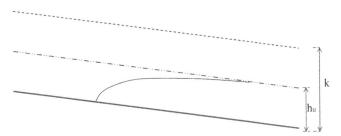

Figure 2.15 Steep slope channel: S3 profile.

2.4 BACKWATER COMPUTATION

As observed in Section 2.3, the water surface profiles can be expressed by equation (2.27), $\frac{dE}{ds} = S_0 - S_f$, once the flow resistance and boundary conditions are known. Equation (2.27) can be written in terms of finite differences as

$$\frac{\Delta E}{\Delta s} = \left(S_0 - \overline{S_f} \right) \qquad (2.50)$$

In order to be able to apply equation (2.50) to compute the water surface profile along the channel, the boundary conditions in the upstream and downstream sections of the control reach are to be carefully determined. In the case of subcritical flows, the control section originating the perturbation of the uniform flow condition is located at the downstream end of the control reach; in supercritical flows, the control section acts from upstream to downstream, thus it is located in the upstream end of the control reach. Thus, when computing the backwater effect in subcritical conditions, the known boundary condition (starting point) will be the downstream known value of the water depth at the control section; whereas in supercritical conditions, the known boundary condition (starting point) for the water surface profile computation will be at the upstream end of the reach, where the control point is located.

In this paragraph, we refer to the backwater computation in subcritical flow conditions. In this case, the starting point for the water surface profile is the downstream control section. The flow is forced by a downstream control section to increase the water depth; going upstream, the flow depth will tend to the uniform flow condition (see M1 and M2 profiles). Thus, both the downstream (water depth at the downstream control section, h^*) and upstream (normal depth, h_u) water depth are known. Let us consider the case of a bridge over a mild slope channel, acting as a downstream control section: at the bridge site, the water depth increases, producing a backwater effect which propagates upstream until reaching the uniform flow condition (assuming that no other control sections will disturb the water profile). Figure 2.16 shows the expected water profile for this example.

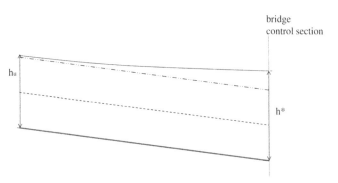

Figure 2.16 Example of water surface profile: backwater effect due to a bridge over a mild slope channel.

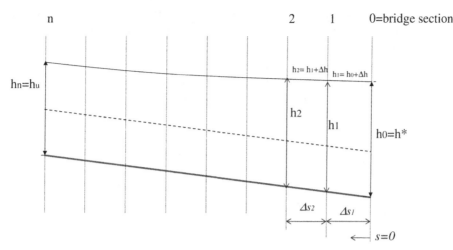

Figure 2.17 Sketch for computing water surface profile using the standard step method.

The difference in water depth between the downstream and upstream sections is then known. We can divide the profile between the two sections into n intervals, going from the downstream bridge section (section 0) to the upstream uniform condition section (section n). The spacing of the intervals does not need to be constant! A finer spacing of the intervals next to the downstream section is suggested for a better representation of the backwater effect due to the presence of the control section. In each of the n sections we can compute the water depth as the sum of the downstream water depth and the increment Δh. Thus, the only unknown in equation (2.50) is the Δs, the distance of the new upstream section from the known downstream section (Figure 2.17). By inverting equation (2.50), one gets:

$$\Delta s = \frac{\Delta E}{(S_0 - \overline{S_f})} \qquad (2.51)$$

With reference to Figure 2.17, to apply equation (2.51), one starts from the downstream known water depth (section 0) and selects the new water depth h_1 (section 1), greater than the downstream water depth h_0, and computes Δs_1, which is the distance of section 1 from section 0. ΔE is the variation in the specific energy along the interval Δs_1 and can be easily computed as

$$\Delta E = E_1 - E_0 = \left(h_1 + \frac{Q^2}{2gA(h_1)^2}\right) - \left(h_0 + \frac{Q^2}{2gA(h_0)^2}\right) \qquad (2.52)$$

Since the specific energy is a function of the water depth only, ΔE can be easily computed once the piezometric head, h, and the kinetic head, $\frac{Q^2}{2gA(h)^2}$, have been calculated in the downstream and upstream sections.

S_0 is the bed slope and $\overline{S_f}$ is the mean friction slope in the Δs_1 interval:

$$\overline{S_f} = \left(\frac{S_{f1} + S_{f0}}{2}\right) \qquad (2.53)$$

To compute the mean friction slope in the interval, one needs to compute the friction slope in section 0 and section 1, given by h_0 and h_1. By applying the Manning–Chezy equation, the friction slope can be estimated as

$$S_{f0} = \frac{n^2 Q^2}{A(h_0)^2 R(h_0)^{4/3}} \qquad (2.54)$$

$$S_{f1} = \frac{n^2 Q^2}{A(h_1)^2 R(h_1)^{4/3}} \qquad (2.55)$$

with n Manning's coefficient, $A(h)$ the cross-sectional area and $R(h)$ the hydraulic radius in the section.

Thus, once h_1 has been estimated, by adding a Δh to the known downstream water depth h_0 and all the variable functions of h_1 and h_0 have been calculated, equation (2.31) can be applied to estimate the distance Δs_1 of the selected h_1 water depth from the downstream section. The procedure is repeated upstream for each Δs until the known upstream value h_n is reached.

2.5 EXERCISES

2.1 A discharge $Q = 20$ m^3 s^{-1} has to be conveyed in an open channel with bed slope $S_0 = 0.0008$. Compute the mean velocity, V_0, and normal depth, h_0, for the uniform flow condition when the channel is:

a. Concrete ($n = 0.014$ m$^{1/3}$ s^{-1}), rectangular (width $b = 5$ m) channel

b. Earth ($n = 0.025$ m$^{1/3}$ s^{-1}) trapezoidal (bottom width $b = 3.5$ m; bank slope 1:1) channel

c. Concrete ($n = 0.014$ m$^{1/3}$ s^{-1}) trapezoidal (bottom width $b = 2.5$ m; bank slope 1:1) channel

2.2 Compute the critical depth for a discharge $Q = 8$ m^3 s^{-1} flowing in a concrete ($n = 0.014$ m$^{1/3}$ s^{-1}) trapezoidal channel (bottom width $b = 2.5$ m; bank slope 1:1).

2.3 The trapezoidal cross-section channel reported in Figure 2.18 is characterized by Manning's coefficient $n = 0.015$ m$^{1/3}$ s^{-1} and has to convey the discharge $Q = 100$ m^3 s^{-1}. Compute and plot the specific energy graph for $Q =$ constant.

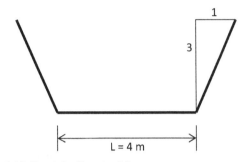

Figure 2.18 Sketch for Exercise 2.3.

2.4 A discharge $Q = 28.0$ (m^3 s^{-1}) is conveyed by a trapezoidal channel. The bottom width of the channel is $b = 7.0$ m, side slope is $m = 1.5$ and Manning's coefficient is $n = 0.025$ (m$^{-1/3}$ s). The channel has a constant bed slope $S_0 = 0.0010$ (mild slope) and ends with a sudden drop of its bed at point B where the slope becomes steep.

a. What type of water surface profile would you expect in the reach A–B? Draw a qualitative profile using Figure 2.19. (Section A is located far upstream of section B.)

b. Compute the profile using the DIRECT STEP method.

c. Plot the water surface profile between A and B.

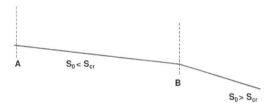

Figure 2.19 Sketch for Exercise 2.4.

3 Theoretical background: unsteady flow

Contributing author: Ioana Popescu

> I shall posit that the fluid cannot be compressed into a
> smaller space, and its continuity cannot be interrupted.
> I stipulate without qualification that, in the course of the
> motion within the fluid, no empty space is left by the fluid,
> but it always maintains continuity in this motion.
>
> Euler (*Principia Motus Fluidorum*, 1756)

3.1 INTRODUCTION

Many flow phenomena are unsteady in nature and cannot be reduced to a steady flow formulation. The most important unsteady flow phenomenon that an engineer has to deal with is the movement of a flood wave downstream of a river channel, often referred to as unsteady flow in open channels.

In general, unsteady flow is a continuum of fluctuations, while unsteady flow in open channels deals with discrete disturbances with long wavelengths. Consequently, the equations describing unsteady flow in open channels are sometimes referred to as 'long wave equations' (Witham, 1974). The long waves are the ones for which the ratio (flow depth)/(wavelength) is much smaller than unity.

The unsteady flow phenomena concerned with short waves are, for example, the wind-induced waves over a reservoir. The size of these waves is a function of wind velocity and fetch (distance over which the wind has blown) determined by various empirical formulae. The short waves seldom appear in open channel problems and they are not addressed in this section of the book.

In the case of a flood wave the wavelength depends on the length of the rainfall period and the amount of runoff becoming inflow into the channel and the amount of baseflow coming from groundwater. From case to case, depending on the above-mentioned conditions, it is possible that the flood wavelength is longer than the channel in which the flood propagates.

In order to describe mathematically the unsteady flow in open channels, equations from continuum mechanics are used. Phenomena in continuum mechanics are usually described using six fundamental equations: the *continuity equation*, based on the conservation of mass; the *momentum equations* along the three orthogonal directions of the Euclidean space (derived from Newton's second law of motion); the *thermal energy equation* (obtained from the first law of thermodynamics); and the *equation of state* (an empirical relation between fluid pressure, density and temperature).

Unsteady flow in open channels does not require the thermal energy equation and the first law of thermodynamics and therefore can be solved by the continuity equation and by the momentum equations, assuming that both density and temperature are constant. The obtained equations are the general Navier–Stokes equations for fluid flow, which can be further simplified, under the assumption of shallow water phenomena, to obtain the Saint-Venant equations for open channel flow. Initially Saint-Venant, in 1871, described the open flow by one-dimensional equations of mass conservation (continuity) and conservation of momentum. Since then the subject has been extensively developed and the results have generated a lot of textbooks and monographs, such as Cunge *et al.* (1980).

In the present chapter the general three-dimensional (3D) Navier–Stokes flow equations are presented in Section 3.2, followed by the Saint-Venant equations in Section 3.3. Solutions of the Saint-Venant equations, for different simplified forms of the momentum equation, are presented in Sections 3.4–3.6. Conclusions are presented in Section 3.7. The chapter ends with exercises on the topic.

For further consideration, notations and conventions are first introduced. In the 3D Euclidian space R^3 a system in Cartesian coordinates is considered. The quantities considered in this system of coordinates are: scalars (for which notation in normal or italic fonts is used); and vectors (for which notation in bold letters is used); vectors have components on all three spatial dimensions (x, y, z). Two-dimensional matrices are represented with capital bold letters.

3.2 NAVIER–STOKES EQUATIONS

Equations of motion for fluids were defined by Euler as early as 1756 in *Principia Motus Fluidorum*. Euler mentions that if an

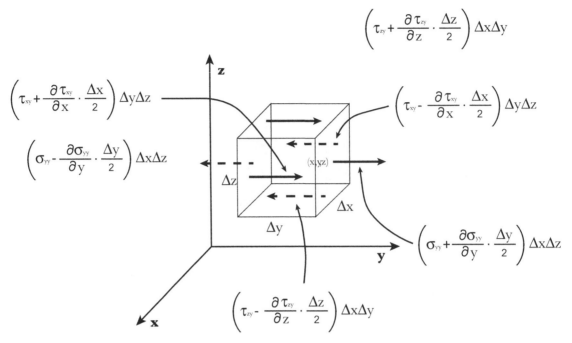

Figure 3.1 Control volume ΔV around the point (x, y, z).

incompressible fluid is taken into consideration then individual particles each fill the same amount of space as they move around. Further, Euler concludes that if this happens for particles it should happen to the fluid as a whole. Therefore, if an arbitrary fluid element is studied in order to see its changes 'to determine the new portion of space in which it will be contained after a very small time period', then the space that is occupied by the fluid element after it moves with time should be equal in size to the old size of the fluid. Euler states, 'This equating of size will fully characterize what can be said about the motion.' This description forms the basis for the so-called control volume method, which is applied further for the elaboration of the equations for fluid motion.

As mentioned, the fluid flow can be expressed mathematically by the use of four equations. The first equation, referred to as the continuity equation, is a mass balance, requiring that the mass of fluid entering a fixed control volume either leaves that volume or accumulates within the volume. The obtained equation is a scalar equation. The other three equations used to describe the fluid flow are the three forms of the momentum equation, on the three directions of space (x, y, z), which represents a 'momentum balance', being the equivalent of Newton's second law (force equals mass times acceleration). The momentum equations form a vector equation. There are many methods to derive these equations. Below, the differential forms of the continuity and momentum conservation laws for incompressible flows are derived, using the control volume approach. An infinitesimally small control volume, ΔV, around the point (x, y, z) is considered

(Figure 3.1) and each term in the equations is expressed for this control volume.

3.2.1 Continuity equation

The mass of fluid in the control volume ΔV depends on the amount of fluid entering and leaving through the faces. The difference between the inflow and the outflow volume is the rate of change in mass or, more simply, it is the mass that accumulates in the volume. The rate of mass entering a face is the product of the density, the fluid velocity and the face area. For example for the area $\Delta y \Delta z$ the mass flux in is

$$mass\ flux\ in = \rho u \Delta y \Delta z \qquad (3.1a)$$

where ρ is the density of the fluid, u is velocity in the x direction and $\Delta y \Delta z$ is the area of that face. The mass flux leaving the control volume in the opposite direction is

$$mass\ flux\ out = -\left(\rho + \frac{\partial \rho}{\partial x}\right)\left(u + \frac{\partial u}{\partial x}\right)\Delta y \Delta z \qquad (3.2a)$$

where density and velocity have changed as the fluid passed through the volume. These changes are small, because the control volume is considered very small, therefore the density and velocity on the opposite face are $\rho + \frac{\partial \rho}{\partial x}$ and $u + \frac{\partial u}{\partial x}$ respectively. The surface area of the face does not change. Expression (3.1a) has a positive sign, since the flux is going into the control volume, while expression (3.2a) has a negative sign, because the flow is going out of the control volume. By analogy, in a similar manner, the expressions for the mass entering and leaving the control volume through the other faces are

$$\rho v \Delta x \Delta z \tag{3.1b}$$

$$\rho w \Delta x \Delta y \tag{3.1c}$$

$$-\left(\rho + \frac{\partial \rho}{\partial y}\right)\left(v + \frac{\partial v}{\partial y}\right)\Delta x \Delta z \tag{3.2b}$$

$$-\left(\rho + \frac{\partial \rho}{\partial z}\right)\left(w + \frac{\partial w}{\partial z}\right)\Delta x \Delta y \tag{3.2c}$$

where v and w are the velocity components in the y and z directions, respectively. The mass of volume accumulating in the control volume can be expressed as

$$\left(\frac{\partial \rho}{\partial t}\right)\Delta x \Delta y \Delta z \tag{3.3}$$

Balancing the accumulated volume (equation 3.3) with the difference between flow in (equation 3.1) and flow out (equation 3.2), the following expression is obtained:

$$\left(\frac{\partial \rho}{\partial t}\right)\Delta x \Delta y \Delta z = \rho u \Delta y \Delta z + \rho v \Delta x \Delta z + \rho w \Delta x \Delta y$$
$$-\left(\rho + \frac{\partial p}{\partial x}\right)\left(u + \frac{\partial u}{\partial x}\right)\Delta y \Delta z$$
$$-\left(\rho + \frac{\partial \rho}{\partial y}\right)\left(v + \frac{\partial v}{\partial y}\right)\Delta x \Delta z$$
$$-\left(\rho + \frac{\partial \rho}{\partial z}\right)\left(w + \frac{\partial w}{\partial z}\right)\Delta y \Delta x \tag{3.4}$$

Dividing by $\Delta x \Delta y \Delta z$ and neglecting all terms that are products of small quantities, such as $\frac{\partial \rho}{\partial x}\frac{\partial u}{\partial x}$ (all higher-order terms), yields

$$\frac{\partial \rho}{\partial t} + \frac{\partial(\rho u)}{\partial x} + \frac{\partial(\rho v)}{\partial y} + \frac{\partial(\rho w)}{\partial z} = 0 \tag{3.5}$$

In the case of flood routing problems, the fluid is considered to be incompressible, and therefore $\rho = ct$. The derivative of ρ with respect to time is zero in this case and equation (3.5) becomes

$$\frac{\partial u}{\partial x} + \frac{\partial v}{\partial y} + \frac{\partial w}{\partial z} = 0 \tag{3.6}$$

Equation (3.6) is scalar and is known as the continuity equation.

3.2.2 Momentum equation

According to Newton's second law of motion:

$$\sum \mathbf{F}_B + \sum \mathbf{F}_S = m\mathbf{a} \tag{3.7}$$

where $\mathbf{F}_B$ and $\mathbf{F}_S$ are the vectors of the external body and surface forces, respectively. The expression $m\mathbf{a}$ represents the vector of the inertia forces (see convention in the introduction, Section 3.1, on how vectors are notated). Mass is represented by the symbol m. The acceleration in a point, represented by $\mathbf{a}$, is the time derivative of the velocity vector $\mathbf{V}(u, v, w)$, with u, v, w the components of the velocity vector in the x, y, z direction:

$$\mathbf{a} = \frac{D\mathbf{V}}{Dt} = \underbrace{\frac{\partial \mathbf{V}}{\partial t}}_{Local\,Acceleration} + \underbrace{u\frac{\partial \mathbf{V}}{\partial x}}_{Convective\,Acceleration} + v\frac{\partial \mathbf{V}}{\partial y} + w\frac{\partial \mathbf{V}}{\partial z} \tag{3.8}$$

The symbol $\frac{D}{Dt}$ indicates the rate of change of the acceleration and it shows that this change is not only dependent on time but it depends on the space variables as well. The components of the acceleration are the local and convection acceleration (see equation 3.8)

Forces over small elements are defined in terms of stresses (force over area). In a moving viscous fluid, forces act not only normal to a surface but also tangential to it. Normal stresses are represented by σ and shear stresses are represented by τ. In Figure 3.1 each stress component has two subscripts, the direction in which it is oriented and the area over which it is acting. For example σ_{xx} represents a force in the x direction that is also acting over an area that is in the x direction. An area of a control volume is considered positive or negative, depending on whether the outward normal to the area points in the positive or negative coordinate direction. Stresses are assumed all positive (i.e. forces act on positive faces in the positive direction, or on negative faces in the negative direction).

Based on the stresses acting over the considered control volume, $\Delta V = \Delta x \cdot \Delta y \cdot \Delta z$ (Figure 3.1), the surface forces that act, for example in the y direction, can be expressed as

$$\Delta F_{Sy} = \left(\sigma_{yy} + \frac{\partial y_{yy}}{\partial y}\cdot\frac{\Delta y}{2}\right)\Delta x \Delta z$$
$$-\left(\sigma_{yy} - \frac{\partial \sigma_{yy}}{\partial y}\cdot\frac{\Delta y}{2}\right)\Delta x \Delta z$$
$$+\left(\tau_{xy} + \frac{\partial \tau_{xy}}{\partial x}\cdot\frac{\Delta x}{2}\right)\Delta y \Delta z$$
$$+\left(\tau_{xy} - \frac{\partial \tau_{xy}}{\partial x}\cdot\frac{\Delta x}{2}\right)\Delta y \Delta z$$
$$+\left(\tau_{zy} + \frac{\partial \tau_{zy}}{\partial z}\cdot\frac{\Delta z}{2}\right)\Delta x \Delta y$$
$$-\left(\tau_{zy} - \frac{\partial \tau_{zy}}{\partial z}\cdot\frac{\Delta z}{2}\right)\Delta x \Delta y$$
$$= \left(\frac{\partial \tau_{xy}}{\partial x} + \frac{\partial \sigma_{yy}}{\partial y} + \frac{\partial \tau_{zy}}{\partial z}\right)\Delta x \Delta y \Delta z$$
$$= 2\cdot\frac{\partial \sigma_{yy}}{\partial y}\cdot\frac{\Delta y}{2}\cdot\Delta x \Delta z + 2\cdot\frac{\partial \tau_{xy}}{\partial x}\cdot\frac{\Delta x}{2}\cdot\Delta y \Delta z$$
$$+2\cdot\frac{\partial \tau_{zy}}{\partial z}\cdot\frac{\Delta y}{2}\cdot\Delta x \Delta z \tag{3.9}$$

Similarly, the body forces on the y face can be expressed as

$$\Delta F_{By} = \rho B_y \cdot \Delta x \Delta y \Delta z \tag{3.10}$$

where B_y are body force components, such as gravity, and the assumption that the fluid is incompressible ($\rho = ct$) still stands.

With expressions (3.9) and (3.10), equation (3.7), for the y direction, can be written as

$$\sum F_{Sy} + \sum F_{By} = m\alpha_y \qquad (3.11)$$

The mass of the small control volume is $\rho\Delta V = \rho\Delta x\Delta y\Delta z$, and the y component of the acceleration, according to (3.8) is

$$\alpha_y = \frac{Dv}{Dt} = \frac{\partial v}{\partial t} + u\frac{\partial v}{\partial x} + v\frac{\partial v}{\partial y} + w\frac{\partial v}{\partial z}$$

Thus, equation (3.11) yields (after division with $\Delta x\Delta y\Delta z$ on both sides):

$$\left(\frac{\partial \tau_{xy}}{\partial x} + \frac{\partial \sigma_{yy}}{\partial y} + \frac{\partial \tau_{zy}}{\partial z}\right) + \rho B_y$$
$$= \rho\left(\frac{\partial v}{\partial t} + u\frac{\partial v}{\partial x} + v\frac{\partial v}{\partial y} + w\frac{\partial v}{\partial z}\right) \qquad (3.12)$$

The first three terms on the left hand side of equation (3.12) represent surface forces per unit volume and the fourth term is the body force per unit volume. The inertia force per unit volume is represented on the right hand side of equation (3.12).

By analogy, similar expressions for the x and z directions can be written, obtaining finally the following three equations:

$$\left(\frac{\partial \sigma_{xx}}{\partial x} + \frac{\partial \tau_{yx}}{\partial y} + \frac{\partial \tau_{zx}}{\partial z}\right) + \rho B_x$$
$$= \rho\left(\frac{\partial u}{\partial t} + u\frac{\partial u}{\partial x} + v\frac{\partial u}{\partial y} + w\frac{\partial u}{\partial z}\right) \qquad (3.13a)$$

$$\left(\frac{\partial \tau_{xy}}{\partial x} + \frac{\partial \sigma_{yy}}{\partial y} + \frac{\partial \tau_{zy}}{\partial z}\right) + \rho B_y$$
$$= \rho\left(\frac{\partial v}{\partial t} + u\frac{\partial v}{\partial x} + v\frac{\partial v}{\partial y} + w\frac{\partial v}{\partial z}\right) \qquad (3.13b)$$

$$\left(\frac{\partial \tau_{xz}}{\partial x} + \frac{\partial \tau_{yz}}{\partial y} + \frac{\partial \sigma \tau_{zz}}{\partial z}\right) + \rho B_z$$
$$= \rho\left(\frac{\partial w}{\partial t} + u\frac{\partial w}{\partial x} + v\frac{\partial w}{\partial y} + w\frac{\partial w}{\partial z}\right) \qquad (3.13c)$$

The above three equations are known as the Navier equations. The analysis of these equations shows that they have 4 independent variables (x, y, z, t) and 12 dependent variables (the components of the velocity vector and the stress components). It is assumed that the body force components are known, as they are generally gravity, electromagnetic, centrifugal and Coriolis forces. As such, there are nine more unknowns than the number of equations, therefore the Navier equations cannot be solved. Stokes proposed a set of constitutive relations, given below, which made these equations solvable. The Stokes constitutive relations, in the defined Cartesian coordinates (Figure 3.1), are

$$\tau_{xy} = \tau_{yx} = \mu\left(\frac{\partial v}{\partial x} + \frac{\partial u}{\partial y}\right)$$
$$\tau_{yz} = \tau_{zy} = \mu\left(\frac{\partial w}{\partial y} + \frac{\partial v}{\partial z}\right) \qquad (3.14a)$$
$$\tau_{zx} = \tau_{xz} = \mu\left(\frac{\partial u}{\partial z} + \frac{\partial w}{\partial x}\right)$$

$$\sigma_{xx} = -\rho - \frac{2}{3}\mu\nabla\cdot\mathbf{V} + 2\mu\frac{\partial u}{\partial x}$$
$$\sigma_{yy} = -\rho - \frac{2}{3}\mu\nabla\cdot\mathbf{V} + 2\mu\frac{\partial v}{\partial y} \qquad (3.14b)$$
$$\sigma_{zz} = -\rho - \frac{2}{3}\mu\nabla\cdot\mathbf{V} + 2\mu\frac{\partial w}{\partial z}$$

where μ is the dynamic viscosity.

The Navier equations, together with the continuity equation, and with the Stokes relations make the Navier equations solvable and are generally known as the Navier–Stokes system of equations; for incompressible flows, in three dimensions (the 3D Navier–Stokes equations for fluid flow):

$$\begin{cases} \nabla\cdot\mathbf{V} = \dfrac{\partial u}{\partial x} + \dfrac{\partial v}{\partial y} + \dfrac{\partial w}{\partial z} = 0 \\[2mm] \rho\left(\dfrac{\partial u}{\partial t} + u\dfrac{\partial u}{\partial x} + v\dfrac{\partial u}{\partial y} + w\dfrac{\partial u}{\partial z}\right) \\ \quad = \rho B_x - \dfrac{\partial p}{\partial x} + \mu\left(\dfrac{\partial^2 u}{\partial x^2} + \dfrac{\partial^2 u}{\partial y^2} + \dfrac{\partial^2 u}{\partial z^2}\right) \\[2mm] \rho\left(\dfrac{\partial v}{\partial t} + u\dfrac{\partial v}{\partial x} + v\dfrac{\partial v}{\partial y} + w\dfrac{\partial v}{\partial z}\right) \\ \quad = \rho B_y - \dfrac{\partial p}{\partial y} + \mu\left(\dfrac{\partial^2 v}{\partial x^2} + \dfrac{\partial^2 v}{\partial y^2} + \dfrac{\partial^2 v}{\partial z^2}\right) \\[2mm] \rho\left(\dfrac{\partial w}{\partial t} + u\dfrac{\partial w}{\partial x} + v\dfrac{\partial w}{\partial y} + w\dfrac{\partial w}{\partial z}\right) \\ \quad = \rho B_z - \dfrac{\partial p}{\partial z} + \mu\left(\dfrac{\partial^2 w}{\partial x^2} + \dfrac{\partial^2 w}{\partial y^2} + \dfrac{\partial^2 w}{\partial z^2}\right) \end{cases} \qquad (3.15)$$

Often the continuity equation and the incompressible Navier–Stokes equations are written in vector form as

$$\begin{cases} \nabla\cdot\mathbf{V} = 0 \\ \rho\dfrac{D\mathbf{V}}{Dt} = \rho B - \nabla p + \mu\nabla^2\mathbf{V} \end{cases} \qquad (3.16)$$

where p is the pressure.

The first equation of the system (3.16) is a scalar equation, while the second equation contains, in a single expression, three equations (one for each of the x, y and z components). Thus there are four equations and four unknowns. The second equation in (3.16), is the vector form of Newton's second law of motion and, while the left hand side contains the inertia force per unit volume, each of the three terms on the right hand side represents a type of external force per unit volume (gravity, pressure and viscous force, respectively).

3.3 SAINT-VENANT EQUATIONS

The classical Saint-Venant system of equations for describing fluid flow can be introduced using physical arguments, but it can also be derived from the 3D free surface incompressible Navier–Stokes equations (3.16) using the classical shallow water assumption, when a first-order approximation is considered and a vertical

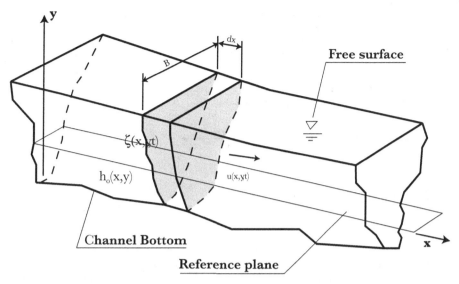

Figure 3.2 Free surface flow representation, for a channel of arbitrary cross section.

integration is applied (Audusse and Bristeau, 2007; Gerbeau and Perthame, 2001).

The basic assumptions for the analytical derivation of the Saint-Venant equations (Chow *et al.*, 1998) are the following:

1. the flow is 1D, i.e. the velocity is uniform over the cross section and the water level across the section is represented by a horizontal line;
2. the streamline curvature is small and the vertical accelerations are negligible, so that the pressure can be taken as hydrostatic;
3. the longitudinal axis of the channel is approximated as a straight line;
4. the average channel bed slope is small so that the cosine of the angle it makes with the horizontal may be replaced by unity, i.e. the effects of scour and deposition are negligible;
5. the effects of boundary friction and turbulence can be accounted for through resistance laws analogous to those used for steady-state flow, i.e. relationships such as the Manning and Chezy equations which relate velocity, hydraulic radius (area divided by wetted perimeter), slope and friction coefficient can be used to describe resistance effects;
6. the fluid is incompressible.

These assumptions do not impose any restriction on the shape of the cross section of the channel and on its variation along the channel axis, although assumption 4 is limited by the condition of small streamline curvature.

In order to present the Saint-Venant equations, a 2D Cartesian reference system is introduced and a control volume of a fluid is represented in this system of coordinates (Figure 3.2). The figure shows the water level (of an open channel flow situation) with

respect to a datum and the control volume between cross sections located at the distance Δx from each other.

The integral form of the Saint-Venant equations, in two dimensions, as they are derived from the 3D Navier–Stokes equations, taking into consideration the basic assumptions of shallow water equations, is (Audusse, 2005):

$$\frac{d}{dt} \int_\Omega \mathbf{u}(\mathbf{x}, t) d\Omega + \oint_\Gamma (\mathbf{f} n_x + \mathbf{g} n_y) d\Gamma$$
$$= \int_\Omega \mathbf{s}(\mathbf{x}, t) d\Omega \, \forall t \in [0.T] \qquad (3.17)$$

where Ω is any open subset of $\mathbb{R}^2$ with boundary Γ; $\mathbf{n}$ is the outward unit normal; and the vectors are

$$\mathbf{u} = (h_z q_x, q_y)^T; \mathbf{f} = \left(q_x, \frac{q_x^2}{h} + \frac{g}{2} h^2, \frac{q_x q_y}{h} \right)^T;$$
$$\mathbf{g} = \left(q_x, \frac{q_x q_y}{h} + \frac{q_y^2}{h} + \frac{g}{2} h^2 \right)^T; \mathbf{s} = \left(0, gh \frac{\partial h_o}{\partial x}, gh \frac{\partial h_o}{\partial y} \right)^T$$
$$(3.18)$$

The components of the vectors (3.18) are: the unit-width discharge, $q(x, t)$, with its components q_x and q_y on the x, y directions; the depth under the reference plane in Figure 3.2, $h_o(x, y)$; the elevation of the water surface above the reference plane, $\zeta(x, y, t)$; the gravitational acceleration, g; the source term that accounts for the bottom slope, $\mathbf{s}$; and $h(x, y, t) = h_o + \zeta$.

Equations (3.18) are used in differential forms as well:

$$\frac{\partial \mathbf{u}}{\partial t} + \frac{\partial \mathbf{f}}{\partial x} + \frac{\partial \mathbf{g}}{\partial y} = \mathbf{s} \quad \text{or} \quad \frac{\partial \mathbf{u}}{\partial t} + \mathbf{B} \frac{\partial \mathbf{u}}{\partial x} + \mathbf{C} \frac{\partial \mathbf{u}}{\partial y} = \mathbf{s}$$
$$\text{with} \quad \mathbf{B} = \frac{\partial \mathbf{f}}{\partial \mathbf{u}} \quad \text{and} \quad \mathbf{C} = \frac{\partial \mathbf{g}}{\partial \mathbf{u}} \qquad (3.19)$$

where $\mathbf{B}$ and $\mathbf{C}$ are the Jacobian matrices of the fluxes $\mathbf{f}$ and $\mathbf{g}$ respectively.

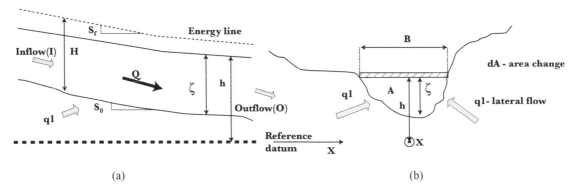

Figure 3.3 River longitudinal profile (a) and arbitrary cross section (b).

Equations (3.19) are the conservative form of the Saint-Venant equations. If the bottom of the channel is flat (i.e. $h_o = 0$) the right hand side of the equation is zero and the obtained equation is a strong conservative form of the Saint-Venant equations.

It is worthwhile mentioning that the Saint-Venant equations have many conservative forms expressing the conservation of mass, energy, discharge rate, velocity, etc. (Ambrosi, 1995). Any two of the conservative forms are equivalent to one another, except for the case when shocks (bores) are involved.

Before the advent of high-speed computers, the Saint-Venant equations were difficult to solve, because of the non-linearity of the momentum equation, and therefore seldom taken into consideration in their form (3.18) or (3.19). However, routing of flow was an important task for engineers, therefore different methods, of varying complexity, were developed for flow routing and especially for determining the flood peak arrival and value, at a certain point of interest. The simplest methods used to determine flood flow through an open channel were the lumped flow routing methods, known as hydrologic methods, which do not use the Saint-Venant equations directly. Nowadays, engineers do solve the Saint-Venant equations with methods known as hydraulic routing methods. The available hydraulic models depend on the desired accuracy and are based on the mass conservation equation (except the lateral flow, in some cases) and several simplifications of the momentum equation.

The three main classical hydraulic models, based on the number of terms considered in the momentum equation, are: kinematic, where the only terms of the momentum equation are the friction and bed slope terms; diffusion, where the momentum equation consists of three terms, those from the kinematic case plus the pressure term; and fully dynamic, which is described by the full momentum equation. A fourth hydraulic routing model (the gravity model) was introduced by Ponce and Simons (1977), where the momentum equation contains inertia and pressure terms. The gravity model is not described in this chapter and the reader interested in it can find further details in the above-mentioned reference.

The demonstration of a solution approach for the three main classical hydraulic models is now explained for the Saint-Venant equations written in 1D form, because of its simplicity. The approach can be then extended to the 2D form of the equations.

From equation (3.19), the conservative 1D representation of the Saint-Venant equations, for a channel of arbitrary cross section A, is

$$\frac{\partial A}{\partial t} + \frac{\partial Q}{\partial x} = q \qquad (3.20a)$$

$$\frac{\partial Q}{\partial t} + \frac{2Q}{A}\frac{\partial Q}{\partial x} + \left(-\frac{Q^2}{A^2} + \frac{gA}{B}\right)\frac{\partial A}{\partial x} - gAS_0 + gAS_f = qu_q \qquad (3.20b)$$

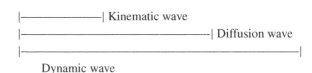

Dynamic wave

where A is the area of the cross section, B is the width of the cross section at the water surface, S_0 is the bottom slope of the channel (positive for a downward channel), S_f represents the friction slope, Q the discharge, q the lateral flow, and u_q the velocity of the lateral inflow (Figure 3.3). In equation (3.20) the three main forms of the momentum equation are shown. The solution approach for each of the above-mentioned hydraulic models is detailed in the following sections of the chapter.

3.4 KINEMATIC WAVE MODEL

As can be seen from equation (3.20b), in the case of the kinematic wave, the acceleration and pressure terms in the momentum equation are neglected, the remaining terms in the momentum equation representing the steady uniform flow. Though the momentum conservation of flow is steady, the effects of unsteadiness are taken into consideration through the continuity equation. In the case of zero lateral flow ($q = 0$), analytical solutions of the differential equation (3.20a) can be derived.

In the case of the kinematic wave, the mathematical model is described by the following set of equations:

$$\begin{cases} \dfrac{\partial A}{\partial t} + \dfrac{\partial Q}{\partial x} = q \\ S_f = S_0 \end{cases} \qquad (3.21)$$

Applying classical differentiation rules to the mass conservation equation, the first equation of the system (3.21), yields

$$\frac{\partial A}{\partial Q}\frac{\partial Q}{\partial t} + \frac{\partial Q}{\partial x} = q \qquad (3.22)$$

which is a form of the well-known advection equation, with the wave celerity $c = \frac{\partial A}{\partial Q}$.

The equation is solvable if the wave celerity, c, can be expressed explicitly. An expression for the kinematic wave celerity may be obtained from one of the steady flow equations such as Chezy or Manning (Chow *et al.*, 1998). Manning's equation states that

$$u = \frac{R^{2/3}}{n}S_f^{1/2} \qquad (3.23)$$

where u is average velocity, R is hydraulic radius ($= A/P$) of the cross section of the channel, P is the wetted perimeter, and n is Manning's coefficient. Given that $Q = Au$ and $S_0 = S_f$, Manning's equation can be rewritten:

$$A = \frac{Q}{u} = \left[\frac{nP^{2/3}}{S_0^{1/2}}\right]Q^{3/5} \quad \text{or in general } A = \alpha Q^\beta \qquad (3.24)$$

Substituting (3.24) in the first equation of the system (3.21) the mass conservation equation yields

$$\frac{\partial Q}{\partial x}\alpha\beta Q^{\beta-1}\frac{\partial Q}{\partial t} = q \qquad (3.25)$$

From basic differential operations $dQ = \frac{\partial Q}{\partial x}dx - \frac{\partial Q}{\partial t}dt$, which, after dividing by dx, yields the volume of flow per unit length:

$$\frac{dQ}{dx} = \frac{\partial Q}{\partial x} + \frac{\partial Q}{\partial t}\frac{dt}{dx} \qquad (3.26)$$

By comparing (3.25) with (3.22), the conclusion is that they are equivalent, provided that

$$\frac{dQ}{dx} = q \quad \text{and} \quad \alpha\beta Q^{\beta-1} = \frac{dt}{dx} \qquad (3.27)$$

Because $\alpha\beta Q^{\beta-1}$ is $\frac{dA}{dQ}$, then the kinematic wave velocity can be computed as $c = \frac{dx}{dt} = \frac{\partial Q}{\partial A}$, and

$$\frac{\partial Q}{\partial x} + c\frac{\partial Q}{\partial x} = cq \qquad (3.28)$$

Equation (3.28) is known as the kinematic wave equation.

Ponce and Simon (1977) analysed the kinematic wave equation, assuming a Chezy formula for the discharge, and indicated that kinematic waves are propagating downstream only; they present no attenuation and their celerity is equal to 1.5 times the mean flow velocity. If Manning's formula is used instead of Chezy's, then a different value of the wave celerity is obtained. The propagation of a wave upstream (backwater effect) is usually computed through the local acceleration, convective acceleration and the pressure terms, all of which are neglected in a kinematic wave approach. When such effects need to be determined, the diffusive or fully dynamic model should be used.

3.5 DIFFUSIVE MODEL

Because the kinematic wave model does not attenuate the flood wave and shows just the translation of the flood wave, it is important to take into consideration the pressure term of the momentum equation. This leads to the diffusive model. Mathematically the model is expressed by the following system of equations:

$$\begin{cases} \dfrac{\partial A}{\partial t} + \dfrac{\partial Q}{\partial x} = q \\ \left(\dfrac{gA}{B} - \dfrac{Q^2}{A^2}\right)\dfrac{\partial A}{\partial x} + g(S_f - S_0)qu = 0 \end{cases} \qquad (3.29)$$

The term 'diffusion' comes from Fick's law in physics, and states that the flux of material along a channel, with cross-sectional area A, is proportional to the gradient of concentration along the channel. This implies the dilution, until it fades out, of the concentration of a solid, liquid, gas or energy. Mathematically, diffusion is represented by an expression where the rate of change of the concentration, with respect to time, is proportional to the second derivative with respect to distance. The diffusive equation is obtained by combining the two equations of the system (3.29) into a single equation of the form:

$$\frac{\partial Q}{\partial t} + c\frac{\partial Q}{\partial x} + D\frac{\partial^2 Q}{\partial x^2} = 0 \qquad (3.30)$$

where c is the kinematic wave celerity (as previously determined) and D is the diffusion coeficient. This expression is obtained by differentiating the continuity equation with respect to distance and substituting it in the momentum equation. In the case where B is constant, and a Chezy approach for expressing the discharge is used, equation (3.30) becomes

$$\frac{\partial Q}{\partial t} + c\frac{\partial Q}{\partial x} = \frac{Q}{2BS_0}\frac{\partial^2 Q}{\partial x^2} \qquad (3.31)$$

The obtained equation is called the convection–diffusion equation and describes the advection of material directly with the flow, while there is a diffusion of material from a higher concentration to a lower concentration.

There are several methods for computing the solution of equation (3.31), such as the Muskingum–Cunge method (Cunge, 1969), the zero-inertial solution (Strelkoff and Katapodes, 1977) or parabolic and backwater (Todini and Bossi, 1986). The method of solution detailed here is the Muskingum–Cunge method, which is one of the most frequently used. The Muskingum–Cunge method uses an equation relating the computation of the discharge at a moment in time as a recursive relation from the values of the discharge computed in the previous time step, i.e.

$$Q_{i+1}^{n+1} = C_1 Q_i^{n+1} + C_2 Q_i^n + C_3 Q_{i+1}^n C_0 \qquad (3.32)$$

where i represents the point in space where the discharge is computed and n represents the point in time. At times n, all values of the discharge are known, and at time $(n + 1)$, all the values of the discharge are computed along the x direction of flow. Coefficients C_i are defined as

$$C_1 = \frac{\Delta t + 2KX}{2K(1 - X) + \Delta t} \qquad (3.33a)$$

$$C_2 = \frac{\Delta t - 2KX}{2K(1 - X) + \Delta t} \qquad (3.33b)$$

$$C_3 = \frac{2K(1 - X) - \Delta t}{2K(1 - X) + \Delta t} \qquad (3.33c)$$

$$C_0 = \frac{q_i \Delta x \Delta t}{2K(1 - X) + \Delta t} \qquad (3.33d)$$

where K is defined as a storage constant with dimensions of time t, and X as a weighting factor showing the relative importance of inflow and outflow to the storage in the control volume. Their mathematical expressions are

$$K = \frac{\Delta x}{c} \quad \text{and} \quad X = 0.5 \left(1 - \frac{Q_0}{cBS_f \Delta x} \right) \qquad (3.34)$$

where Q_0 is the mean discharge, c is the kinematic wave speed, B is the width of the water at the top of the cross section, equivalent to Q_0, S_f is the friction slope, and Δx is the reach length.

It can be quickly checked that the following relation stands:

$$C_1 + C_2 + C_3 = 1C_1, \quad C_3 > 0 \qquad (3.35)$$

Equation (3.32) can be solved by either a linear or a non-linear method. For the linear solution, K and X are assumed constant for all time steps in each point in space, on the x direction.

The non-linear solution is more accurate. An initial estimated value of Q_{i+1}^{n+1} and the water elevation, corresponding to this value of the discharge, are used to calculate K and X. Then the solution is computed iteratively until there is convergence (accepted small value differences between the assumed h and the computed one). For numerical stability (Cunge, 1969), constraints have to be imposed on the values of K and X, in the form:

$$0 \leq \Delta x \leq \frac{1}{2} \quad \text{and} \quad 2X < \frac{\Delta t}{K} < 2(1 - x) \qquad (3.36)$$

In the case of the diffusion model there is only one wave propagated in the downstream direction of the flow, limiting as such the use of the method for cases where strong backwater effects appear.

3.6 FULLY DYNAMIC MODEL

The fully dynamic model uses the full momentum equation, because it is considered that the inertial forces are as important as the pressure forces, hence it uses the full representation of the Saint-Venant equations (3.30). These equations can be solved only by using numerical methods for solving partial differential equations, making use of computer power to solve the system of equations obtained after the application of different numerical schemes to equations (3.30).

In the literature, several numerical techniques for solving the Saint-Venant equations are known. These include the method of characteristics, explicit difference methods, semi-implicit methods (Casulli, 1990), fully implicit methods and Godunov methods (van Leer, 1979). The characteristic method transforms the Saint-Venant partial differential equations into a set of ordinary differential equations. In this way, four ordinary differential equation are obtained and solved, easily, using finite difference methods.

The explicit methods transform the Saint-Venant equations into a set of algebraic equations, expressed at each point of discretization in space and time. These equations are solved, one at a time. The implicit methods solve the obtained algebraic equations simultaneously at all computational points at a specific moment in time. Iteration is needed sometimes due to boundary conditions and non-linearity. Solving the Saint-Venant equations numerically poses problems of stability and convergence of the numerical solution. Explicit methods are the ones that introduce errors that may accumulate and pose problems to the stability of the solution. In order to prevent error propagation, the Courant–Frederichs–Levy (CFL) condition is imposed ($\Delta t \leq \Delta x / u$). Implicit methods are subject to constraints in the selection of the time and space steps of computation, in order to obtain convergence. However, they do not induce stability problems in their solution.

Godunov-type methods are explicit in time, and thus restricted by the CFL stability condition. The methods were originally developed for gas dynamics and later extended to hydrodynamics (Toro, 1997; LeVeque, 2002).

Semi-implicit methods can be unconditionally stable and still computationally efficient. A semi-implicit method that conserves the fluid volume, applied to the case of channels with arbitrary cross sections, was introduced and presented in Casulli and Zanolli (2002).

When, in a semi-implicit scheme, the efficiency of staggered grids is combined with the conservation of both fluid volume and momentum, then problems addressing rapidly varying flow can be solved (Stelling and Duijnmayer, 2003).

For the exemplification of how numerical schemes are implemented to solve the Saint-Venant equations, the four-point Preissmann scheme is explained below. An (x, t) domain of computation of length L and span time T is selected (Figure 3.4). The domain

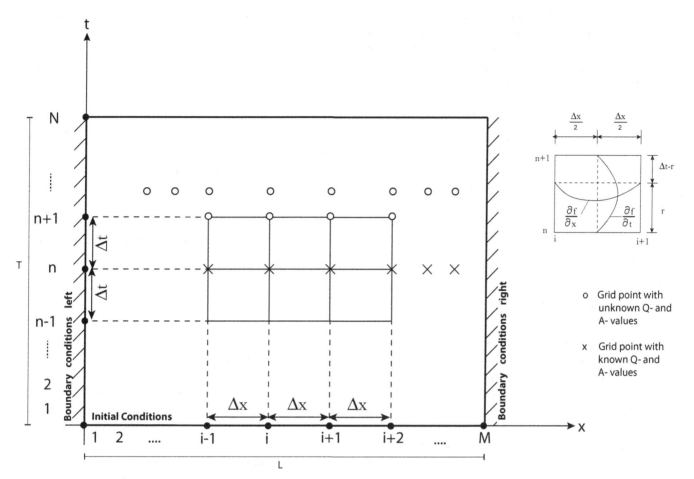

Figure 3.4 Computational domain.

is split M times on Δx spaces of variable length, and N times on equal Δt time intervals. Since the Saint-Venant equations are valid on the entire computational domain, they are valid on the discretization points ($i\,\Delta x$, $n\,\Delta t$) as well. Four points of the domain are shown in Figure 3.4.

In the Preissmann scheme, the time derivative of any dependent variable f is approximated by a forward difference ratio at a point centred between the ith and $(i + 1)$th point along the distance axis and nth and $(n + 1)$th point along the time axis so that

$$\frac{\partial f}{\partial x} = \frac{0.5(f_{i+1}^{n+1} - f_i^{n+1})}{\Delta t} + \frac{0.5(f_{i+1}^{n} - f_i^{n})}{\Delta t} \quad (3.37)$$

The derivative with respect to space is approximated between two adjacent time lines, as follows:

$$\frac{\partial f}{\partial x} = \frac{\theta(f_{i+1}^{n+1} - f_i^{n+1})}{\Delta x} + \frac{(1 - \theta)(f_{i+1}^{n} - f_i^{n})}{\Delta x} \quad (3.38)$$

where θ is the weighting coefficient.

The non-derivative terms of the Saint-Venant equations are approximated (at the same point, using the same coefficients) as

a derivative with respect to x, hence,

$$f = \frac{\theta(f^{n+1} - f_{i+1}^{n+1})}{2} + \frac{(1 - \theta)(f_i^{n} - f_{i+1}^{n})}{2} \quad (3.39)$$

The constraint of the scheme is that θ is usually greater than 0.5.

Substituting these operators into the Saint-Venant equations, for both Q and A, leads to a set of weighted four-point implicit finite difference equations. For a discretization of the computational domain with M points along the x axis, the number of finite difference equations applied to each of the $M - 1$ grid points gives $2M - 2$ equations, and there are in total $2M$ unknowns. In order to make the system determined and solvable, boundary conditions at the upstream and downstream end are required. Solution of the equations is usually done using the double sweep algorithm (Abbott and Basco, 1989).

There are many different numerical schemes encapsulated in different software codes, such as: the weighted six-point Abbott–Ionescu scheme (Abbott and Ionescu, 1967), the weighted four-point Preissmann scheme (Preissmann, 1961), alternate direction implicit (Stelling and Duijnmaijer, 2003) and TVD (total variation diminishing) schemes (Toro, 1997). Each numerical scheme has its own advantages and disadvantages. The fully implicit schemes of Preissmann and Abbott–Ionescu were the first to be used in hydrodynamic computational codes, in the early 1960s. Though these codes have been developed a lot since then, in terms of the graphical user interface, the main numerical schemes remain the initial ones, and are still widely used nowadays. Two examples of codes that use the Preissmann scheme are DAMBRK, which was developed by the US National Weather Service, and FLUVIAL, which was developed by the University of Illinois. An example of code that uses the Abbott–Ionescu scheme is Mike11, developed at the Danish Hydraulic Institute. Godunov-based schemes with various Riemann solvers are used in river modelling software such as Infoworks RS 2D (Roca and Davison, 2009), TRENT (Villanueva and Wright, 2006) and BreZo (Begnudelli *et al.*, 2008).

3.7 CONCLUSIONS

This chapter has presented the basic theory of unsteady flow for the case of free surface flow, which is of interest in flood propagation problems. Flood propagation downstream of a channel, also referred to as flood routing, is a phenomenon described by open channel flow in natural channels, which due to its free surface is non-uniform and unsteady in nature. The unsteadiness introduces the need to express two dependent variables (discharge and depth, or velocity and depth) as functions of space and time. Depending on the dimension of space, taken into consideration for expressing the basic flow equations, the independent variables are x (for 1D computations), or x and y (for 2D computations), or all three space directions x, y and z.

The full representation of the governing equations of the free surface flow, either in differential form or in integral form, is called the Saint-Venant system of equations or the dynamic wave equations. There are very few analytical solutions available for these equations, even under major simplifications. Nevertheless, there are two major simplifications of these equations, which were presented in this chapter. The applicability of these is described below.

The first, and most simple, simplification of the fully dynamic wave equation is the kinematic wave approximation, which does not give an exact model for the movement of a flood wave. Consequently this model can be applied to a limited range of flood problems (Lighthill and Whitham, 1955). The kinematic wave is used for cases where the rating curve does not present loops and there are insignificant backwater effects. The slope of the channels for which the kinematic wave can be applied should be from moderate to steep, where hydrograph

propagation does not present strong attenuation from upstream to downstream.

The kinematic wave approximation normally does not show attenuation, unless the solution is computed using a numerical approximation, which induces numerical errors known as diffusion errors.

The second simplified model is the diffusive wave model, which has a wider applicability than the kinematic wave model, but still is limited to situations where backwater effects are not significant.

In the case of rivers where inertia terms are important, the fully dynamic equations should be used for the computation of flood propagation downstream of a river system. Moreover, taking into account today's computer power, there is no real reason not to work with the fully dynamic wave equation. The use of the three forms of the equation is the choice of the engineer solving a flood propagation problem, and it depends on knowledge of the particular river system to be solved, rather than saving time of computation.

3.8 EXERCISES

3.1. In a river reach, an inflow hydrograph has a peak discharge of 1,000 $\text{m}^3 \text{ s}^{-1}$. The hydrograph is triangular in shape and has a baseflow of 300 $\text{m}^3 \text{ s}^{-1}$. The time to peak is 6 hours, and it takes another 6 hours to recess back to the baseflow. The channel has a length of 10 km, a slope of 0.0001 and a rectangular cross section of 500 m width. Manning's coefficient n of 0.025 should be considered for computation. Find the outflow peak discharge and the time of occurrence by applying the Muskingum–Cunge method for 24 hours. Consider 6 hours of baseflow situation before the hydrograph is applied at the upstream end of the channel.

3.2. Route the hydrograph of Exercise 3.1 by applying the kinematic wave routing method. Compare the obtained results.

3.3. Route the hydrograph of Exercise 3.1 in a channel with the same geometrical characteristics and the same length, but with a different slope, $S_0 = 0.002$. Compare and comment on the obtained results.

3.4. Derive the kinematic wave celerity for the cases of a rectangular and a trapezoidal channel. In the case of a trapezoidal channel plot the ratio c/u (where c is the kinematic wave velocity and u is the average channel flow velocity) as a function of the aspect ratio b/h for several values of side slope ratio m. Comment the result.

3.5. Determine the algebraic form of the Saint-Venant equations if the Preissmann scheme is applied to them on a domain of length L and time span of computation T.

Part II
Methods

4 Data sources

Flood propagation and inundation modelling strongly relies on the available data (Cunge, 2003). More specifically, numerical hydraulic models require: (i) topographical data for the description of the geometry of rivers and floodplains, (ii) hydrometric observations or rating curves for the definition of initial and boundary conditions, and (iii) hydrographs, high flood marks or flood extent maps for model calibration, validation and uncertainty analysis.

This chapter discusses these different data sources by describing both traditional ground-surveyed data (e.g. cross sections, hydrometric data) and remotely sensed data (e.g. satellite and airborne images). The chapter focuses more on the latter as the potential given by the recent diffusion of remote sensing data, which has led to a sudden shift from a data-sparse to a data-rich environment for flood inundation modelling (Bates, 2004a), is still not entirely utilized by most hydraulic modellers and flood managers.

4.1 GROUND DATA

4.1.1 Topography

The most common types of field data used for the topographical description of rivers and floodplains are planimetric maps, river profiles, geometrical description of hydraulic structures and cross sections. Ground-surveyed cross sections, in particular, are often used as geometrical input of hydraulic models. Choosing a suitable set of cross sections for the representation of the natural geometry of rivers and floodplains is therefore important for the efficiency of hydraulic models.

However, the identification of the optimal topographic survey in hydraulic modelling (i.e. the selection of the optimal number of cross sections and their best location) is usually performed following subjective criteria. For instance, it is unlikely that two hydraulic modellers would select exactly the same cross-section location (Samuels, 1990). Intuitively, on the one hand the larger the number of cross sections, the better the performance of the model. On the other hand, the overall cost of the

topographic survey undoubtedly increases with the number of cross sections.

Guidelines for selection of the most suitable distance between cross sections, depending on the hydraulic problem at hand, were reported in Cunge et al. (1980) and Samuels (1990) and then tested by Castellarin et al. (2009). Aside from the obvious recommendations (i.e. cross sections should be located: (i) at the model upstream and downstream ends; (ii) at either side of structures where an internal boundary is set; (iii) at each point of specific interest; (iv) at all available stream gauges), various other suggestions were given. A recommended distance between cross sections, Δx, was

$$\Delta x \approx k B \qquad (4.1)$$

where B is the bankfull surface width of the main channel and k is a constant (with a recommended range from 10 to 20). This first guideline expresses the intuitive argument that larger Δx should be used for larger rivers.

Based on an estimate of backwater effects for subcritical flows, it was suggested that

$$\Delta x < 0.2 \frac{(1 - Fr^2) D}{s} \approx 0.2 \frac{D}{s} \quad \text{when } Fr^2 \to 0 \qquad (4.2)$$

where D is the bankfull depth of flow and s is the surface (or main channel) slope. Over this length, the backwater upstream of a control (as well as other disturbance) decays to less than 0.1 of the original value.

For unsteady conditions, an accurate representation of physical waves may be desirable. Two different waves may be of importance, the flood wave and the tide wave propagating along an estuary. Samuels (1990) pointed out that a reasonable representation of the physical wave requires a number of grid points N_{gp} between 30 and 50, and therefore recommends that

$$\Delta x < \frac{cT}{N_{gp}} \qquad (4.3)$$

where the wavelength is expressed as the product of the period T and the propagation speed c of the wave and N_{gp} varies between 30 and 50.

Finally, the effect of rounding error can also be taken into account, identifying a minimum distance between cross sections as

$$\Delta x > \frac{10^{d-q}}{s\varepsilon_s} \qquad (4.4)$$

in which q is the number of decimal digits of precision, d represents the digits lost due to cancellation of the leading digits of the stage values, s is the average surface slope, and ε_s is the relative error on surface slope that can be tolerated in the computation. For example, if $d = 2$ (e.g. stage values in cm), $q = 6$, $s \approx 10^{-3}$ and $\varepsilon_s \approx 10^{-3}$ then $\Delta x > 100$ m.

As mentioned above, the equations (4.1)–(4.4) were first proposed by Cunge *et al.* (1980) and Samuels (1990) and then verified through an extensive numerical study by Castellarin *et al.* (2009). Thus, these indications can be used as general guidelines for the identification of the optimal topographic survey in hydraulic modelling.

Another commonly used geometrical input of hydraulic models is digital elevation models (DEMs). The utility of DEMs in flood inundation modelling obviously depends on their accuracy and resolution. Horritt and Bates (2001) evaluated the effects of spatial resolution on hydraulic modelling of flood inundation by testing DEM resolutions varying from 1,000 m to 10 m and comparing model predictions with satellite observations of inundated areas and ground measurements of flood wave travel times. The study showed that the inundation model reached the maximum performance at a resolution of 50 m, after which no improvement was seen with increasing resolution (Horritt and Bates, 2001).

Given that, nowadays, DEMs are often derived using remote sensing techniques, more details about resolution and accuracy of DEMs and their use in flood inundation modelling are included in Section 4.2.1.

4.1.2 Hydrometry

Hydrometric data recorded in gauging stations are often used in flood propagation and inundation modelling both as input (boundary and initial conditions) and as calibration data. In particular, the common implementation of hydraulic models is based on the use of the following hydrometric data: (i) observed river discharges in the upstream end of the modelled river reach, as the upstream boundary condition; (ii) observed water levels (or stage–discharge rating curve) at the downstream end, as the downstream boundary condition; and (iii) observed water levels at internal cross sections, as calibration data. The travel time of the flood peak can also be used for model calibration.

In terms of accuracy, the errors reported in the literature for water level observations are around 2–5 cm (Pappenberger *et al.*, 2006) and therefore often negligible compared to the other sources of uncertainty in flood inundation modelling (Section 4.3). By

contrast, river discharges are almost never directly measured. Usually, observed river stage values are converted into river discharges by means of a stage–discharge relationship, the so-called rating curve (World Meteorological Organisation, 1994). Thus, river discharges are affected by a significant uncertainty (Clarke, 1999), which may be very high for data referred to high flow conditions when stage–discharge rating curves are extrapolated beyond the measurement range (Petersen-Øverleir, 2004). For instance, Di Baldassarre and Montanari (2009) pointed out that uncertainty affecting recorded river discharge data might be as high as 30% of the observed value (Section 4.3).

4.1.3 High flood marks

Another example of ground data that can be used for model calibration is high water marks and wrack marks (Neal *et al.*, 2009a). As an example, Figure 4.1 shows the high water marks (i.e. post-event measured maximum water levels) surveyed in a reach of the River Po (Italy) after the October 2000 flood event (Coratza, 2005). In particular, Figure 4.1 reports the observed maximum water levels at left and right embankments of the river reach. According to the scientific literature, the accuracy of high flood marks can be estimated as between 30 and 50 cm (Neal *et al.*, 2009a; Horritt *et al.*, 2010).

It is worth mentioning here the current development of cheap wireless computing sensors to monitor floodplains and therefore support flood predictions. These sensors (e.g. GridStix; Huges *et al.*, 2007) are able to measure water levels which are then telemetered in real time using GSM phone technology. The real time water levels can then be assimilated into flood routing and flood inundation models to improve forecast performance. These cheap sensors may improve our ability to predict flood inundation modelling in the near future.

4.2 REMOTE SENSING DATA

As mentioned above, the increasing availability of distributed remote sensing data has led to a sudden shift from a data-sparse to a data-rich environment for flood inundation modelling (Alsdorf *et al.*, 2001, 2007; Bates, 2004a). For instance, airborne laser altimetry data produce a wealth of topographic information for inclusion into inundation models (Cobby *et al.*, 2001), while flood extent maps derived from remote sensing are useful calibration data for the evaluation of hydraulic models (Horritt *et al.*, 2007).

4.2.1 Topography

An accurate description of the geometry of rivers and floodplains is crucial for the efficiency of hydraulic models. Modern

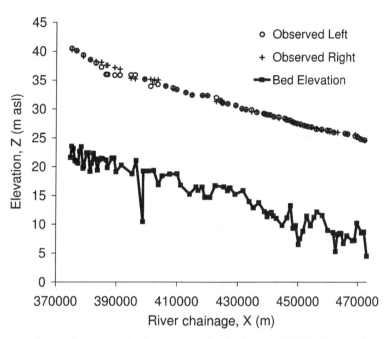

Figure 4.1 Example of high water marks: maximum water levels measured after the October 2000 flood in the River Po (Italy).

techniques for topographical survey (e.g. LiDAR, Cobby *et al.*, 2001) enable numerical description of the morphology of riverbanks and floodplains with planimetric resolution of 1 m and finer. These techniques, along with the increasing availability of GIS (Geographic Information System) tools for hydrologic and hydraulic studies, have expanded the traditional application field for hydraulic models, making them easily suitable for predicting flood inundation extent (Bates, 2004a).

In the past decade, airborne laser altimetry has experienced a great diffusion in western countries. For instance, more than 62% of England and Wales (81% of urban areas) is covered by LiDAR survey (2010). According to the Geomatics Group of the Environment Agency of England and Wales, the elevation accuracy of this LiDAR data is between 5 and 15 cm. Thus, this type of topographic data can allow detailed hydraulic studies and urban inundation modelling.

Figure 4.2 shows, as an example of LiDAR topography, the DEM of a reach of the River Po (Italy), which was built on the basis of the data collected in year 2005 during numerous flights, using two different laser-scanners (3033 Optech ALTM and Toposys Falcon II), from altitudes of approximately 1500 m (Castellarin *et al.*, 2009). Below water, channel bathymetry of the navigable portion was derived by a boat survey using a multi-beam sonar (Kongsberg EM 3000D), conducted in the same year, integrated elsewhere with the information collected during a previous ground survey consisting of traditional cross sections conducted by AIPO (Interregional Authority of the Po River), still in 2005. The resulting DEM (Figure 4.2) was validated against the data achieved through a differential global positioning system

(DGPS). Mean quadratic residuals between DGPS survey and DEM were found to be less than 13 cm for approximately 25,000 control points located over a 150 km reach. Also, the validation procedure confirmed the absence of local systematic differences (Camorani *et al.*, 2006). DEM generation usually involves the removal of surface features such as vegetation from the data set of aggregated heights (Cobby *et al.*, 2001), which is the case in the example reported in Figure 4.2.

In recent years, there has been a great diffusion of topographic data that are freely and globally available, such as the space-borne DEM derived from the SRTM, which has a geometric resolution of 3 arc seconds (around 90 m; e.g. LeFavour and Alsdorf, 2005). Although of low accuracy (around 6 m), this type of data can be extremely useful for large rivers. For instance, a number of studies have demonstrated the potential of SRTM data to derive useful hydraulic parameters, such as water surface slope and discharge (LeFavour and Alsdorf, 2005). More recently, Schumann *et al.* (2010) demonstrated that globally and freely available low-resolution space-borne data sets can be used to approximate the longitudinal profile of a flood wave on larger rivers. In particular, they showed that SRTM-derived water profiles have a significant value for the evaluation of hydraulic models, and are able to discriminate effectively between competing model parameterizations. Confirmation of their utility therefore indicates the potential to remove an important obstacle currently preventing the routine application of hydraulic models to predict flood hazards globally, and potentially allows such technology to be extended to developing countries that have not previously been able to benefit from flood predictions.

Figure 4.2 Example of LiDAR topography: DEM of a reach of the River Po (Italy) at a spatial resolution of 2 m, after the removal of vegetation and buildings (greyscale: black, 50 m a.s.l.; white, 10 m a.s.l.).

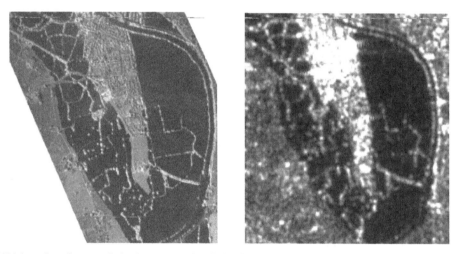

Figure 4.3 Example of high- and medium-resolution imagery used to derive flood extent maps: airborne SAR (left) and RADARSAT (right) imagery of flood inundation at Upton-upon-Severn, UK, in November 2000 (Bates *et al.*, 2006).

4.2.2 Flood extent maps

From space, satellites carrying SAR sensors are particularly useful for monitoring large flood events (Aplin *et al.*, 1999). In fact, radar wavelengths, which can penetrate clouds and acquire data during day and night, are reflected away from the antenna by smooth open water bodies, and hence mapping of water surfaces becomes relatively straightforward. Schumann *et al.* (2009a) have described currently available SAR remote sensing techniques to retrieve flood boundaries and water levels from space-borne imagery and subsequently reviewed the studies that have tried to integrate these with flood inundation

models for more rigorous performance evaluation or uncertainty reduction.

Satellite and airborne imagery used for flood extent mapping can be characterized by different spatial resolutions, which, relative to the typical length scales of physical flow process during floods, can be broadly defined as: high (1–2 m), fine/medium (10–25 m), or coarse/low (about 100 m). Figure 4.3 shows, as an example, a comparison of high- and medium-resolution flood imagery acquired during the November 2000 flood in Upton-upon-Severn, UK (Bates *et al.*, 2006).

Airborne SAR imagery (Figure 4.3) is an example of high-resolution flood imaging, and has been used recently to better understand floodplain inundation processes and compare different hydraulic models with satisfactory results (Wright *et al.*, 2008). In particular, given the accuracy of flood extent maps derived from airborne SAR images (around 2 m; Horritt *et al.*, 2007), they are extremely useful for model evaluation. However, the acquisition of airborne SAR imagery requires the expensive organization of ad-hoc flights. The newly launched SAR sensors onboard TerraSAR-X (11 days revisit time) and RADARSAT-2 (24 days revisit time), which have been recently used to map flooding in urban areas (Mason *et al.*, 2010), are also examples of high-resolution flood imaging. Nevertheless, their costs and relatively long revisit times compared to the duration of typical flooding episodes in the majority of river basins are not appropriate for cost-effective use in everyday practice.

Examples of medium-resolution flood images are those derived by ERS-2 or RADARSAT (Figure 4.3). The effectiveness of this type of image has been tested in a number of model validation studies (Bates and De Roo, 2000; Horritt and Bates, 2001). Nevertheless, competing priorities in mission design mean that orbits have repeat overpass times of 35 days for the ERS-1 and ERS-2 instruments and therefore it is almost impossible to acquire more than one image per flood (Hunter *et al.*, 2007). It is worth noting that RADARSAT's capability of reducing orbital acquisition times considerably, by tilting through a range of different incidence angles, can help overcome temporal constraints.

By contrast, lower-resolution flood images have the advantages of greater spatial and temporal coverage, lower cost and lack of copyright restrictions. For instance, the European Space Agency (ESA) ENVISAT-ASAR (Advanced Synthetic Aperture Radar) wide swath mode (WSM) revisit times can be of the order of 3 days and images can be quickly obtained (data latency of ~24 hours) at no (or low) cost to users. The fact that coarser resolution SAR data can be used successfully to delineate flood edges on large floodplain areas (inundation widths larger than 500 m) has been demonstrated, for example, by Blyth (1997) and Kussul *et al.* (2008) within a grid system, and by the International Disaster Charter (www.disasterscharter.org/). Moreover, a number of studies (Brandimarte *et al.*, 2009; Schumann *et al.*, 2009a; Schumann and Di Baldassarre, 2010) have recently demonstrated the usefulness of coarse-resolution flood images for supporting flood modelling in medium-to-large rivers.

Table 4.1 reports spatial resolution and repeat cycles of current satellite missions featuring SAR sensors with high potential for flood propagation and inundation studies (Schumann *et al.*, 2010).

To derive flood extent maps from airborne or satellite imagery, prior to image processing, a filter (e.g. Sigma Lee filter; Smith, 1996) should be applied to SAR imagery to remove most speckle, i.e. random image noise obstructing features of interest. Then, many different image processing techniques may be applied to a satellite image. However, it is well known that no single method

Table 4.1 *Summary of current missions useful for reconstruction of flood extent maps*

Mission (agency: year of launch)	Spatial resolution (m)	Repeat cycle (days)
ERS-2 (ESA: 1995)	25	35
RADARSAT-1 (CSA: 1995)	8–100	24
ENVISAT (ESA: 2002)	12.5–1000	35
ALOS (JAXA: 2006)	7–100	46
COSMO-SkyMed (ASI: 2007)	15–100	16
TerraSAR-X (DLR: 2007)	1–16	11
RADARSAT-2 (MDA: 2007)	3–100	24

Di Baldassarre *et al.*, 2011a

can be considered appropriate for all images, nor are all methods equally good for a particular type of image (Schumann *et al.*, 2009a). This section provides some notions of some of the most common procedures: visual interpretation, histogram threshold, active contour, and image texture variance. More details can be found in Schumann *et al.* (2009a). Section 4.3.2 illustrates an application of different image processing procedures to the same imagery (Figure 4.5).

There are a number of studies that illustrate the potential of visual interpretation to derive flooded area from satellite imagery (Oberstadler *et al.*, 1997). In this approach, the flooded area is mapped by visually digitizing the flood boundaries. A skilful delineation of flood shorelines by visual interpretation requires expert flood knowledge. If breaks of slope between the floodplain and adjacent hillslopes are clearly visible in the topography, they should be used to constrain the delineated flood extent to the valley floor area.

Histogram thresholding is a simple but widely used and efficient method to generate binary maps from images. An optimal grey-level threshold can be found using the Otsu method (Otsu, 1979). The method applies a criterion measure to evaluate the between-class variance (i.e. separability) of a threshold at a given level computed from a normalized image histogram of grey levels.

The active contour method is based on a dynamic curvilinear contour that searches the edge image space until it settles upon image region boundaries. This is achieved by an energy function attracted to edge points. The contour is usually represented as a series of nodes linked by straight line segments (Horritt *et al.*, 2001). The statistical snake is formulated as an energy minimization. The total energy is minimized if the contour encloses a large area of good pixels, and in this respect the model behaves as a region-growing algorithm (Horritt *et al.*, 2001).

Image texture can be modelled as a grey-level function using simple statistical methods on the image histogram. Widely used algorithms rely on statistical properties of a neighbourhood of pixels that are computed for each pixel using a moving window.

Table 4.2 *Advantages and disadvantages of commonly used image processing techniques to obtain flood extent maps from SAR imagery*

	Visual interpretation	Histogram thresholding	Texture based	Active contour modelling/ Region growing
Strength	Easy to perform in the case of a skilled and experienced operator with knowledge of flood processes	Easy and quick to apply Objective method	Takes account of the SAR textural variation Based on statistics Mimics human interpretation as it takes account of tonal differences	Image statistics based Usually provides good classification results Easy to define seed region (e.g. on the river channel) If integrated with land elevation constraints results are improved by mimicking inundation processes
Limitation	Very subjective Difficult to implement over many images May be difficult for images that show complex flood paths	No flexibility Optimized threshold might not be the most appropriate Works well only if image is relatively little distorted	Difficult to choose correct window size and appropriate texture measure After application still requires threshold value to obtain flood area classification	Requires several parameters to fine-tune Slow on large image domains Difficult to choose correct tolerance criterion May miss separated patches of dry or flooded land
Level of complexity	Low to high (may have varying degrees of complexity)	Very low	Moderate	Moderate to high
Computational efficiency	Relatively low	Very high	Moderate	Moderate (strongly depending on domain size)
Level of automation	Hardly possible	Full	Full	Relatively high
Consistency[a]	0.9	0.8	0.6[b]	0.7

Di Baldassarre *et al.*, 2011a

[a] Refers to consistency of binary classification between different SAR images, after Schumann *et al.* (2009a).

[b] Average of different texture measures.

The image texture variance and mean Euclidean distance (Irons and Petersen, 1981) can be found in most commercial remote sensing software packages.

An indication of the advantages and disadvantages of these image processing techniques to obtain flood inundation extent from SAR imagery is presented in Table 4.2.

4.2.3   Flood water levels

As mentioned above, water levels are typically measured in traditional river gauging stations. Yet water levels can be derived by means of sonar boats equipped with GPS using real time kinematic (RTK) satellite navigation or satellite profiling altimeters (Alsdorf *et al.*, 2007), which are able to cover larger areas. Coherent pairs of radar images can also be processed interferometrically to yield maps of water-level change (Alsdorf *et al.*, 2007). However, this is only possible over flooded vegetation where a double bounce allows a signal to be returned to the sensor.

The extraction of water levels at the flood shoreline can also be performed by intersecting flood extent maps with high-resolution digital terrain models (DTMs), particularly from airborne laser altimetry (Schumann *et al.*, 2007a). Accounting for flood mapping and related uncertainties across entire floodplain sections perpendicular to the flow direction has enabled this technique to be augmented by estimations of uncertainties associated with shoreline heights and approximation of water surface gradients thereof (Schumann *et al.*, 2008, 2010). Table 4.3 lists water-level retrieval techniques based on remotely sensed flood extent and DEM fusion.

4.3   UNCERTAINTY

Table 4.4 reports the most common types of input and calibration data and gives an indication of the associated uncertainty. The flood travel time (i.e. time of propagation of the peak level) is not

Table 4.3 *Water level retrieval techniques based on remotely sensed flood extent and DEM fusion and their accuracies*

Method	Accuracy	Validation data	Source
Landsat TM-derived flood extent superimposed on topographic contours for volume estimation	±21%	Field data	Gupta and Banerji (1985)
ERS-SAR flood extent overlain on topographic contours	0.5–2 m	Field data	Oberstadler *et al.* (1997)
ERS-SAR flood extent overlain on topographic contours	<2 m	Model outputs	Brakenridge *et al.* (1998)
Inter-tidal area shorelines from multiple ERS images superimposed onto simulated shoreline heights	0.2–0.3 m	Field data	Mason *et al.* (2001)
Flooded vegetation maps from combined airborne Land C-SAR integrated with LiDAR vegetation height map	around 0.1 m	Field data	Horritt *et al.* (2003)
Integration of high-resolution elevation data with event wrack lines	<0.2 m	Model outputs	Lane *et al.* (2003)
Fusion of RADARSAT-1 SAR flood edges with LiDAR	Correlation = 0.9	TELEMAC-2D model outputs	Mason *et al.* (2003)
Complex fusion of flood aerial photography and field-based water stages from various floodplain structures	0.23 m	Mean between maximum and minimum estimation	Raclot (2006)
Fusion of ENVISAT ASAR flood edges with LiDAR and interpolation modelling	0.4–0.7 m	Field data	Matgen *et al.* (2007)
Fusion of ENVISAT ASAR flood edges with LiDAR and regression modelling	<0.2 m	Field data	Schumann *et al.* (2007a)
Fusion of ENVISAT ASAR flood edges with LiDAR/topographic contours/SRTM and regression modelling	<0.35 m <0.7 m <1.07m	1D model outputs	Schumann *et al.* (2008)
Fusion of hydraulically sensitive flood zones from ENVISAT ASAR imagery and LiDAR	0.3–0.5 m	Mean between maximum and minimum estimation	Hostache *et al.* (2009)
Fusion of ERS SAR flood edges from active contour modelling with LiDAR	<0.5 m	Aerial photography	Mason *et al.* (2009)
Fusion of TerraSAR-X flood edge with LiDAR (rural area)	1.2 m[a]	One field gauge	Zwenzner and Voigt (2009)
Fusion of uncertain ENVISAT ASAR WSM flood edges with SRTM heights	error in median estimate of 0.8 m	LiDAR derived water levels	Schumann *et al.* (2010)

Di Baldassarre *et al.*, 2011a

[a] Note that relative changes in levels from TerraSAR-X and aerial photography were mapped with an accuracy of 0.35 m compared to gauge data.

Table 4.4 *Summary of data used in flood inundation modelling and indicative order of magnitude of the associated uncertainty*

Data	Common use	Accuracy	Reference
Gauged river discharges	input	20–40%[a]	Di Baldassarre and Montanari (2009)
Gauged water levels	input/calibration	0.02–0.05 m	Pappenberger *et al.* (2006)
High water marks	calibration	0.3–0.5 m	Neal *et al.* (2009a)
Flood extent maps (Airborne SAR)	calibration	1–2 m	Horritt *et al.* (2007)
Flood extent maps (ENVISAT-ASAR WSM)	calibration	150–300 m	Schumann *et al.* (2009a)

[a] Referred to high-flow conditions (floods), when the rating curve is extrapolated beyond the measurement range.

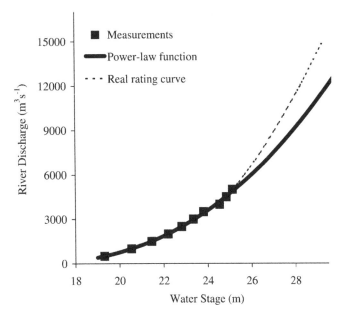

Figure 4.4 Rating curve estimation: interpolation of contemporaneous measurements of water stage and river discharge by means of the power-law function and extrapolation beyond the range of measurements (Di Baldassarre *et al.*, 2011a).

included in the table as its accuracy is equal to the time resolution of the river gauging stations. More details on the accuracy of hydrometric data are reported in Herschy (1978) and the European ISO EN Rule 748 (1997).

Table 4.4 clearly shows that river discharge observations and flood extent maps derived from coarse resolution imagery are affected by a relevant uncertainty. Thus, the next sections focus on these two types of data uncertainty.

4.3.1 Uncertainty in river discharges

As mentioned above, river discharge observations are usually obtained by means of stage–discharge rating curves (Figure 4.4). In particular, the standard methodology to derive a rating curve consists of carrying out field campaigns to record contemporaneous measurements of water stage, h, and river discharge, Q. Such measurements allow one to identify discrete points (Q, h) that are subsequently interpolated through an analytical relationship that approximates the rating curve. The power-law function is commonly used in hydrometric practice (Herschy, 1978):

$$Q = a \cdot (h - b)^c \qquad (4.5)$$

where a, b and c are calibration parameters that are usually estimated by means of the non-linear least squares method (Petersen-Øverleir, 2004). Equation (4.5) is widely used in river hydraulics and has some physical justifications (Petersen-Øverleir, 2004).

Thus, the main sources of uncertainty in river discharge observations are (e.g. Di Baldassarre and Claps, 2011): (i) errors in

individual stage and discharge measurements used to build the rating curve; (ii) errors induced by the presence of unsteady flow conditions; (iii) errors induced by the extrapolation of the rating curve beyond the range of measurements used for its derivation. Depending on the specific case study, additional sources of uncertainty can be significant, such as the presence of relevant backwater effects (caused by downstream confluent tributaries, lakes and regulated reservoirs) and temporal changes in the hydraulic properties governing the stage–discharge relationship (e.g. scour and fill, vegetation growth, ice build-up during cold periods).

Concerning the measurement uncertainty (case i), Pelletier (1987) reviewed 140 publications and concluded that the overall uncertainty in a single determination of river discharge can be more than 8% at the 95% confidence level. More recent studies reported errors around 5–6% (e.g. Léonard *et al.*, 2000), which could be possibly reduced by using appropriate discharge measurement techniques (Lintrup, 1989; European ISO EN Rule 748, 1997).

The errors induced by the presence of unsteady flow (case ii) can be relevant in very mild river slope conditions, where the variable energy slope leads to the formation of a loop rating curve (Jones, 1916). It is worth noting that, to reduce this source of uncertainty, an original approach based on simultaneous stage measurements at two adjacent cross sections was recently proposed (Dottori *et al.*, 2009).

Finally, the uncertainty induced by the extrapolation of the rating curve beyond the measurement range (case iii) can result in an amplification of the previous uncertainties (Figure 4.4). Given the lack of measurements during high-flow conditions, indirect and extrapolated discharge measures of flood discharges turn out to be affected by relevant errors, so that many authors warn not to extrapolate rating curves beyond a certain range (e.g. Rantz *et al.*, 1982). For instance, Di Baldassarre and Montanari (2009) performed a quantitative numerical analysis to estimate the uncertainty of river discharge observations on the River Po (Italy) and showed that the errors produced by the extrapolation of the rating curve beyond the range of measurements used for its derivation were about 14% at the 95% confidence level. They also showed that this extrapolation uncertainty strongly increases for increasing values of the river discharge (Figure 4.4).

Nevertheless, river discharge data referred to high-flow conditions are needed for the hydraulic modelling of floods and floodplain mapping. Thus, the extrapolation of the rating curve beyond the measurement range is very often a necessity and more efforts are needed to reduce the errors and uncertainties associated with this indirect measure (Chapter 10).

Hydraulic modellers and flood managers should therefore bear in mind this source of uncertainty, which is very often neglected. This book provides general guidelines to cope with the uncertainty in river discharge observations in the evaluation of hydraulic models (Chapter 6) and floodplain mapping (Chapter 7). Moreover, a

Figure 4.5 River Dee, UK, in December 2006: medium (ERS-SAR, left) and low (ENVISAT-ASAR, right) resolution SAR imagery (Di Baldassarre *et al.*, 2009b).

strategy to reduce the uncertainty induced by the extrapolation of stage–discharge rating curves is described in Chapter 10.

4.3.2 Uncertainty in space-borne flood extent maps

The observed flood extent maps are commonly treated as deterministic maps when in reality the observed flood extent is subject to considerable uncertainty, especially when derived from low/medium-resolution satellite imagery.

To facilitate the description of the uncertainty in space-borne flood extent maps, this chapter refers to a specific case study: a river reach of the Lower Dee (UK). This test site is of particular interest because, during the December 2006 flood event, both coarse-resolution (ENVISAT ASAR WSM) and medium-resolution (ERS-2 SAR) satellite imagery were acquired at the same time (Figure 4.5; Schumann *et al.*, 2009a). Thus, this unique data set enables a discussion of the uncertainty in flood extent maps derived from satellite imagery as, in terms of the inundation process, the actual flood extent at both acquisitions can be assumed to be the same. Moreover, all apparent differences in flood extent mapping on both images can be attributed to differences in spatial resolution, given that all other significant acquisition parameters (e.g. frequency, polarization and incidence angle) were the same for both data sets (Schumann and Di Baldassarre, 2010).

To investigate the uncertainty in SAR-derived flood extent maps, the two satellite images (Figure 4.5) were processed to derive flood extent maps by using five different image processing procedures: visual interpretation, histogram threshold, active contour, image texture variance and Euclidean distance (Section 4.2).

Hence, ten different flood extent maps were derived from the two flood images. By analysing these maps, Schumann *et al.* (2009a) observed significant differences between the outcomes

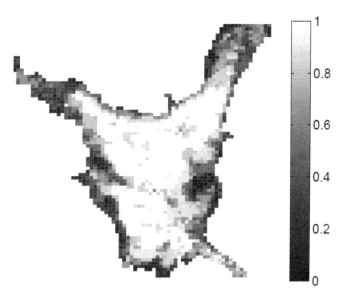

Figure 4.6 Uncertain flood inundation map (right) obtained by combining ten flood extent maps derived from the two imagery types and applying five different image processing techniques (Schumann *et al.*, 2009a).

of different image processing procedures. These differences impacted the use of these flood extent maps as calibration data. For instance, Di Baldassarre *et al.* (2009b) used these ten flood extent maps to calibrate a flood inundation model (LISFLOOD-FP; Bates and De Roo, 2000). The calibration exercise showed that the optimal parameters of the model depend on the type of satellite image used to evaluate the model as well as on the particular image processing technique used to derive the flood extent map. This result clearly demonstrates the necessity to move from traditional, deterministic binary (wet/dry) maps to probabilistic maps of flood extent. To this end, Schumann *et al.* (2009a) produced an uncertain flood inundation map by fusing the ten flood extent maps according to a particular measure of consistency into a single fuzzy flood map.

Figure 4.6 shows this uncertain flood inundation map, where the mapped probability P_j reflects the likelihood of observing inundation at the jth cell for the 2006 flood event. This uncertain flood inundation map may be a useful tool for flood risk mapping, as it expresses our belief as to whether a particular image pixel is flooded by an event of a given magnitude. More details for the derivation of the uncertain flood inundation map can be found in Schumann *et al.* (2009a). A methodology for the calibration of hydraulic models by comparing the model results (with uncertain parameter values) to the uncertain flood inundation map (Figure 4.6) can be found in Di Baldassarre *et al.* (2009b).

4.4 CONCLUSIONS AND PERSPECTIVES

'The numerical hydraulic models are only as good as the data used to calibrate and verify them, so that the new economic utility of the models translates into a new economic utility for field studies and improved instrumentation' (Abbott, 1979). This chapter presented different data sources to support the hydraulic modelling of floods, with a focus on the relatively recent diffusion of remote sensing data. Different data sources and the associated uncertainty were discussed. Particular emphasis was given to the uncertainty in river discharge observations and space-borne flood extent maps because of their potential magnitude and the fact that these sources of uncertainty are often neglected.

Moreover, this chapter discussed the recent demonstration of the value of freely and globally available space-borne data to support hydraulic modelling of larger rivers, which is extremely remarkable. In fact, flood inundation is a global hazard, and the fact that freely and globally available space-borne data have value will substantially increase the number of sites where inundation models can be developed. For instance, they can be used to support flood inundation modelling and floodplain mapping in data-poor areas and developing countries. More specifically, while there has been a wide development of global models simulating climate, weather, large-scale hydrology and flood detection systems (e.g. modelling tools developed by the European Joint Research Centre; the Global Flood Alert System; Cloke and Hannah, 2011; Yamazaki *et al.*, 2011), there is still a lack of global floodplain models able to make worldwide prediction of inundation patterns and therefore identify floodplain areas at a resolution that is useful for floodplain management (25–100 m; Blyth, 1997; Apel *et al.*, 2009). Thus, numerous floodplain systems remain largely unexplored and the identification of flood-prone areas (if any) is often too coarse and/or approximated to effectively support floodplain management (Blyth, 1997; Apel *et al.*, 2009). As a result, detailed floodplain studies are currently available only for a few river reaches and almost non-existent in developing countries (Di Baldassarre *et al.*, 2010a). However, global floodplain inundation models grasping the opportunity given by the current proliferation of globally and freely available data, which has been described as a 'flood of data' (Lincoln, 2007), are still to be developed.

4.5 EXERCISES

4.1. Illustrate the types of data commonly used in flood propagation and inundation modelling.

4.2. Explain why low-resolution imagery, such as ENVISAT-ASAR in WSM, is useful to support hydraulic modelling of floods.

4.3. Illustrate the expected order of magnitude of accuracy of input and calibration data.

4.4. Indicate the main sources of errors affecting river flow data when river discharge values are obtained using stage–discharge rating curves.

4.5. Make a web search, produce an updated list of freely and globally available data that can be used to support flood propagation and inundation modelling, and describe the potential of this information in data-poor areas.

5 Model building

Simplicity is the ultimate sophistication.

Leonardo da Vinci, *circa* 1500

This chapter discusses the issue of model implementation in both theoretical and practical terms. In particular, numerical tools for flood inundation modelling are classified and briefly described. The chapter then introduces the principle of parsimony and the main criteria behind the selection of the most appropriate hydraulic model for simulating flood inundation. Lastly, the chapter deals with many issues related to model building, such as the schematization of model geometry and the parameterization of flow resistance.

5.1 MODELLING APPROACHES

5.1.1 Flood propagation and inundation processes

Lowland rivers usually consist of a main channel and adjacent floodplain areas. When a flood wave exceeds bankfull height, water travels quickly over the low-lying floodplains. During a flood, the floodplain areas may act either as storages or as additional means of conveyance (Woodhead *et al.*, 2009). To select an appropriate flood model, it is important to consider the size of the flood wave. In the largest river basins, waves may be up to around 10^3 km in length, but only around 10 m deep, and may take several weeks or months to traverse the whole system (Bates *et al.*, 2005; Castellarin *et al.*, 2009). Flood waves are translated downstream and attenuated by frictional losses such that in downstream sections the hydrograph is flattened out (Woodhead *et al.*, 2009). Figure 5.1 shows an example of flood propagation between two gauging stations and shows the natural processes of translation and attenuation of the flood wave. The celerity of waves varies with the river discharge. For UK rivers, NERC (1975) and Bates *et al.* (1998) report typical values of celerity between 0.3 m s^{-1} and 1.8 m s^{-1}.

Below the scale of the flood wave, there are other in-channel processes to consider, each with a characteristic length scale (Bates *et al.*, 2005): (i) shear layers forming at the junction between the main flow and slower moving dead zones at the scale

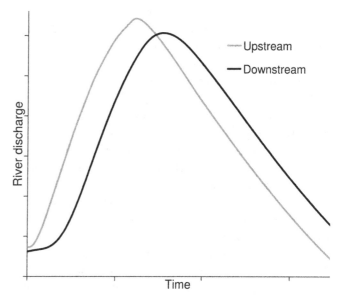

Figure 5.1 Example of translation and attenuation of a flood wave.

of the channel planform (Hankin *et al.*, 2001); (ii) secondary circulations at the scale of the channel cross section (Bridge and Gabel, 1992); and (iii) turbulent eddies ranging from heterogeneous structures at the scale of roughness elements and obstructions on the bed (Ashworth *et al.*, 1996), down through the turbulent energy cascade (Hervouet and Van Haren, 1996), to the Kolmogorov length scale, where turbulent kinetic energy is dissipated. The smallest eddies may be only a few millimetres across, and the grid size required to include such processes in flood inundation models makes this infeasible for most real applications (Woodhead *et al.*, 2009).

When the flow depth exceeds bankfull height, additional physical processes will begin to operate (Bates *et al.*, 2005). The principal mechanisms are momentum exchange between the fast-moving channel and slower floodplain flow (Knight and Shiono, 1996) and interaction between meandering channel flows and flow on the floodplain (Sellin and Willetts, 1996). The channel–floodplain momentum exchange occurs across a shear layer, which is manifest as a series of vortices with vertically aligned axes (Sellin, 1964). Ervine and Baird (1982) indicated that failure to account for the momentum exchange can lead to errors of up to $\pm 25\%$ in the discharge calculated using uniform flow relationships, such as Manning's equations (Woodhead et al., 2009). Further significant momentum exchange occurs during out-of-bank flow in meandering compound channels; more details can be found in Sellin and Willetts (1996).

Away from the near-channel zone, the movement of water on the floodplain may be more accurately described as a typical shallow water flow, as the horizontal extent may be large (up to several kilometres) compared to the depth (usually less than 10 m). Such shallow water flows over flat topography are characterized by rapid extension and retreat of the inundation front over considerable distances, potentially with different processes occurring during the wetting and drying phases (Woodhead et al., 2009). Thus, a proper treatment of this moving boundary problem is essential to capture adequately the shallow water energy losses, which may be high due to relatively high roughness (Bates et al., 2005).

Flow interactions with micro-topography, vegetation and structures may also be important (Bates et al., 2005). In fact, where the floodplain acts as a route for flow conveyance rather than just as a storage, energy losses are typically dominated by vegetative resistance (Woodhead et al., 2009). However, the interactions between plant form, plant biomechanics, energy loss and turbulence generation are still relatively poorly understood (Wilson et al., 2005) and more research is needed.

Furthermore, it should be noted that hydraulic models usually assume that the channel geometry is fixed during the inundation event. This may not be the case for very large floods when embankment failures or geomorphic change may considerably affect the river geometry (Bates et al., 2005). However, during flood events, when river discharges are sufficiently higher than the bankfull discharge, the effects of changes of river geometry in flood levels tend to be minimal in alluvial rivers (Di Baldassarre and Claps, 2011). This is due to the fact that changes in the geometry of the river reach mainly occur in the main channel and therefore do not have a strong effect on the hydraulics of very large floods when the (often stable) floodplain areas give a relevant contribution to the flow conveyance.

Lastly, while many numerical models of floodplain flow do not consider water exchanges with the surrounding catchment, such processes may become important in modelling flood inundation simulation over long river reaches (Woodhead et al., 2009). These processes comprise: direct precipitation or runoff to the floodplain areas; evapotranspiration; the so-called bank-storage effects (Pinder and Sauer, 1971) resulting from interactions between the river water and alluvial groundwater contained within the hyporheic zone; subsurface contributions to the floodplain groundwater from adjacent hill slopes (Bates et al., 2000); and flows along preferential flow paths, such as relict channel gravels, within the floodplain alluvium (Poole et al., 2002). Over particular reaches and in

particular environments, the integration of (some) of these processes with flood routing models may be required and necessitate complex modelling structures (Kohane and Welz, 1994). Nevertheless, it should be noted that these complex modelling exercises often lead to difficulties in the parameterization (Bates et al., 2005).

5.1.2 Classification of models

Nowadays, several numerical tools are available for modelling flood propagation and inundation processes. Pender (2006) classified hydraulic models according to the dimensionality of the represented processes. Table 5.1 reports examples of modelling tools and provides an indication of their potential application; each type of model is appropriate for different tasks and applications over different scales (Pender, 2006).

The hydraulic models reported in Table 5.1 present a variety of required input data, generated outputs and computational costs. Table 5.2 reports these characteristics for different flood inundation models (Pender, 2006; Woodhead et al., 2009).

In addition, a new generation of models able to simulate rapidly varying flood fronts (e.g. Stelling and Duijnmayer, 2003; Liang et al., 2007) as well as advanced tools for modelling the dynamic processes of levee breaches and dam breaks (e.g. Aureli et al., 2008; Savant et al., 2010) are examples of the current developments in computational hydraulics that might further improve our ability to simulate flood propagation and inundation processes.

5.1.3 Modelling tools

Fully 3D Reynolds averaged Navier–Stokes (RANS) models are very expensive for floodplain inundation problems and also do not cope well with dynamic wetting and drying (Di Baldassarre et al., 2010b). Hence the most physically realistic code that is practically applied is a full solution of the 2D shallow water equations (SWE). The 2D approach conserves momentum for the floodplain simulation (Pender, 2006) and numerical solution of the 2D SWE can be obtained from different methods (finite difference, finite volume or finite element; Chapter 3), which utilize different numerical grids (structured or unstructured).

An example of fully 2D model code is TELEMAC-2D (Galland et al., 1991). The TELEMAC-2D code solves the 2D SWE for a system of piecewise linear triangular finite elements using a fractional step method (Marchuk, 1975). The method of characteristics is used for the advection step and the streamline upwind Petrov–Galerkin method (Brookes and Hughes, 1982) is used to solve the combined propagation and diffusion step. The resulting linear system is solved using an element-by-element technique and the generalized minimum residual method. Several studies have shown the capability of TELEMAC-2D to simulate flow over complex topography, dynamic wetting and drying of the

Table 5.1 *Summary of numerical tools for flood inundation modelling and their potential application*

Method	Description	Software examples	Potential application
0D	No physical laws	ArcGIS, Delta mapper	Broad scale assessment of flood extents and flood depths
1D	Solution of the 1D equations	Mike 11, HEC-RAS	Design scale modelling, which can be of the order of tens to hundreds of km depending on catchment size
1D+	1D plus a flood storage cell approach to the simulation of floodplain flow	Mike 11, HEC-RAS	Design scale modelling, which can be of the order of tens to hundreds of km depending on catchment size, also has the potential for broad scale application if used with sparse cross-sectional data
2D−	2D minus the law of conservation of momentum for the floodplain flow	LISFLOOD-FP, CA model	Large-scale modelling or urban inundation depending on cell dimensions
2D	Solution of the 2D shallow wave equations	TUFLOW, Mike 21, TELEMAC, DIVAST	Design scale modelling of the order of tens of km. May have the potential for use in broad scale modelling if applied with coarse grids
2D+	2D plus a solution for vertical velocities using continuity only	TELEMAC 3D	Predominantly coastal modelling applications where 3D velocity profiles are important. Has also been applied to reach scale river modelling problems in research projects
3D	Solution of the 3D Reynolds averaged Navier–Stokes equations	CFX, FLUENT, PHOENIX	Local predictions of 3D velocity fields in main channels and floodplains

Pender (2006)

Table 5.2 *Flood inundation modelling: input data, output and order of magnitude of the computation time*

Method	Input data	Output	Computation time
0D	DEM, Upstream water level, Downstream water level	Inundation extent and water depth by intersecting planar water surface with DEM	Seconds
1D	Surveyed cross sections of channel and floodplain, Upstream discharge hydrographs, Downstream stage hydrographs	Water depth and average velocity at each cross section, Inundation extent by intersecting predicted water depths with DEM, Downstream outflow hydrograph	Minutes
1D+	As 1D models	As 1D models	Minutes to hours
2D-	DEM, Upstream discharge hydrographs, Downstream stage hydrographs	Inundation extent, Water depths, Downstream outflow hydrographs	Hours
2D	DEM, Upstream discharge hydrographs, Downstream stage hydrographs	Inundation extent, Water depths, Depth-averaged velocities at each computational node, Downstream outflow hydrograph	Hours to days
2D+	DEM, Upstream discharge hydrographs, Inlet velocity distribution, Downstream stage hydrographs	Inundation extent, Water depths, Velocity vector at each computational cell, Downstream outflow hydrograph	Days
3D	DEM, Upstream discharge hydrographs, Inlet velocity and turbulent kinetic energy distribution, Downstream stage hydrographs	Inundation extent, Water depths, Velocity vector and turbulent kinetic energy for each computational cell, Downstream outflow hydrograph	Days

Woodhead *et al.* (2009)

floodplain, and mass fluxes between channel and floodplain (e.g. Hervouet and Van Haren, 1996; Horritt and Bates, 2002; Horritt *et al.*, 2007). One of the advantages of finite element models, such as TELEMAC-2D, is that they are based on unstructured meshes that can be used for better describing the topographical discontinuities that influence the inundation process, such as levees, road and railway embankments (Aronica *et al.*, 1998; Di Baldassarre *et al.*, 2009c).

In recent years, many flood inundation models of reduced complexity have been developed. Some of them are based on hybrid schemes that combine 1D modelling for channel flows with a 2D treatment of the floodplain (Bates *et al.*, 2005). In such an approach, main channel flow is modelled using a 1D kinematic or diffusive wave solution (Chapter 3). During out-of-bank flow, water is transferred to a 2D floodplain grid across which a 2D dynamic simulation is undertaken using a Manning-type equation to compute flows between grid cells (Cunge *et al.*, 1980). The concept is similar to the one adopted for the 1D+ approach, but with grid dimensions being considerably smaller than storage areas (Pender, 2006). These hybrid techniques were developed, originally, to take advantage of high-resolution topographic data sets (Bates and De Roo, 2000). An example of hybrid model code is the simple raster-based model LISFLOOD-FP (Bates and De Roo, 2000; Horritt and Bates, 2002). In LISFLOOD-FP, the channel is discretized as a single vector along its centre line separate from the overlying floodplain DTM. At each point along the vector the required channel parameters are the width, Manning's coefficient and bed elevation. The latter data give the bed slope and also the bankfull depth when the channel vector is combined with the floodplain DTM. Each channel parameter can be specified at each point along the vector and the model interpolates linearly between these. Flows along the channel can be simulated using either the diffusive or kinematic approximation to the Saint-Venant equations (Chapter 3), and when the bankfull depth is exceeded water spills out onto adjacent floodplain areas. Floodplain flows are treated using a storage cell approach (Cunge *et al.*, 1980) and implemented for a raster grid to allow an approximation to a 2D diffusive wave. To improve the efficiency of the model, Bates *et al.* (2010) recently developed a simple inertial formulation.

Another interesting example of a reduced complexity model is the recently developed 2D model based on the cellular automata (CA) approach (Dottori and Todini, 2011), which presents the advantage of allowing for the use of irregular meshes.

Many practical floodplain management issues only require the prediction of water levels at particular points of interest (Woodhead *et al.*, 2009). In such cases, the modeller is primarily concerned with flood propagation, and may be less concerned to accurately simulate floodplain flow and storage. In this case, the flow processes of interest are 1D in the down-valley direction and 1D models may therefore be used to represent such flows

(Bates *et al.*, 2005). This simplified approach can be justified by assuming that the additional inaccuracies introduced by treating the out-of-bank flow as if it were 1D are small compared to other sources of uncertainty in hydraulic modelling (Chapter 6).

Widely used software for 1D hydraulic modelling is HEC-RAS (Hydrologic Engineering Center, 2001), which solves the well-known Saint-Venant equations for unsteady open channel flow (Chapter 3) through the UNET code (Barkau, 1997). The equations are discretized using the finite difference method and solved using a four-point implicit method (box scheme; Priessmann, 1961). HEC-RAS can also be used in a 1D+ approach, where floodplain areas are modelled as storage reservoirs, whose geometry is defined using a water level versus volume relationships (e.g. Castellarin *et al.*, 2011). In this approach, floodplain water level is linked to the main channel using spill units that model the flow between the river and the storage areas, using weir flow based discharge relationships (Di Baldassarre *et al.*, 2009c). Water level in each storage area is then computed using conservation of volume (Pender, 2006).

Given the wide diffusion of HEC-RAS (e.g. its use has become common practice in floodplain mapping in the United States; Merwade *et al.*, 2008), many related tools have been developed. Among them, it is worth mentioning HEC-GeoRAS (Ackerman, 2002), which is a tool that works within a GIS environment to preprocess data and postprocess end results associated with the HEC-RAS model.

The electronic resources of this book, which are available online, include the HEC-RAS software and its user manuals. Also, Chapter 6 and Chapter 10 report numerical exercises carried out with HEC-RAS. The interested reader might also check on the website of the US Army Corps of Engineers for possible updated version of the software.

5.2 MODEL SELECTION

5.2.1 Selection criteria

Some decades ago, the famous statistician George Box pointed out, 'All models are wrong, but some are useful'. In fact, although a model can never be perfect, models can be ranked, depending on the specific application, as very useful, somewhat useful or essentially useless (Burnham and Anderson, 2002). This ranking exercise is the essence of model selection (Laio *et al.*, 2009). In general, as the dimension (or complexity) of a model increases, the bias tends to decrease, whereas the uncertainty tends to increase (Figure 5.2; principle of parsimony; Box and Jenkins, 1970). The principle of parsimony is also known as Occam's razor: to remove all that is not needed (Wagenmakers, 2003). Modellers should aim at building parsimonious models that achieve a proper trade-off between bias and uncertainty (Di Baldassarre

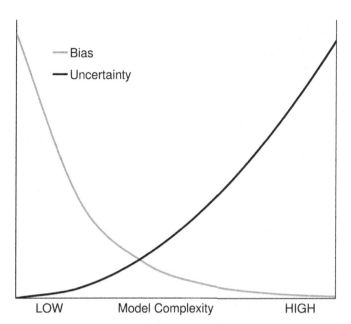

Figure 5.2 The principle of parsimony: the conceptual trade-off between bias (grey) and uncertainty (black) versus model complexity.

et al., 2009d). This also reflects the famous sentence, commonly attributed to Albert Einstein, 'Everything should be made as simple as possible, but no simpler'.

More specifically, an unparsimonious model tends to capture relatively much of the noise in the data (i.e. idiosyncratic information; Wagenmakers, 2003). In fact, by simply adding parameters, it is possible to fit almost everything (Burnham and Anderson, 2002). However, although an unparsimonious model is expected to fit the available data very well, such a model might make poor predictions as its parameter estimates tend to be affected by a relatively high uncertainty (Figure 5.2). Hence, one should avoid using unparsimonious models. Obviously, a model that captures relatively little structural information (or an under-fitted model) is also not well suited.

The art of modelling is based on building parsimonious models that are able to capture the dominant processes that come into play. Figure 5.3 provides a simple example to illustrate this concept. In particular, the top panel of Figure 5.3 shows a natural floodplain where large changes in flood inundation extent are caused by small changes in water levels; here, the dynamics are essentially 2D and the inundation processes are dominated by the floodplain topography. A different example is reported in the bottom panel of Figure 5.3, showing a protected floodplain where the flood shoreline is constrained by manmade embankments. In this other case, the dynamics are principally 1D and the inundation processes are dominated by the presence of manmade embankments (as long as they are not overtopped).

Complex fully 2D models are theoretically able to cope with both options of Figure 5.3. However, complex models might prove difficult to apply because of computational costs, requirement of

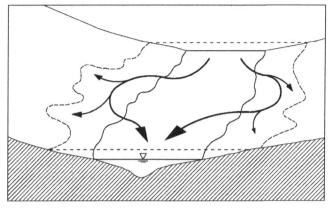

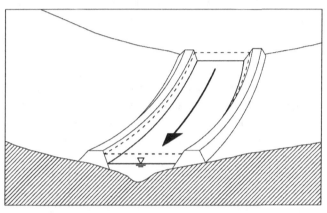

Figure 5.3 Simple schematization of floodplain inundation processes. Top: Example of natural floodplain where small changes in water levels (dotted line) produce large changes in flood inundation extent. In this case, the dynamics are essentially 2D (black arrows) and the inundation processes are dominated by the floodplain topography. Bottom: Example of defended floodplain where the flood shoreline (dotted line) is constrained by manmade embankments. In this case, the dynamics are principally 1D (black arrow) and the inundation processes are dominated by topographic discontinuities, i.e. the manmade embankments. (Sketch kindly drawn and provided by Domenico Di Baldassarre.)

a large amount of high-quality data, and difficulties in parameterization (Bates and De Roo, 2000). By contrast, a simple and parsimonious 1D model (Table 5.1) could make appropriate predictions for the defended floodplain (Figure 5.3, bottom panel), but it might fail in predicting the (essentially 2D) inundation dynamics over a natural floodplain (Figure 5.3, top panel).

In addition, another aspect to take into account in flood inundation modelling is the computational efficiency of the hydraulic model. The selection of an appropriate numerical modelling structure for floodplain flows should be based on the identification of the processes that are most relevant to a particular problem and the assessment that these processes can be discretized and parameterized in a computationally efficient way (Bates *et al.*, 2005). This criterion forces the modellers to think about the minimum process representation needed to make appropriate predictions of particular quantities (Hunter *et al.*, 2007).

Lastly, model selection should also be based on the actual availability (and quality) of data (Chapter 4) and the purpose of the model. For instance, there is no point in using complex models in data-poor areas as our models 'are only as good as the data used to calibrate and verify them' (Abbott, 1979). Moreover, while flood forecasting does need fast models (Szöllösi-Nagy and Mekis, 1988), mapping flood hazard and risk requires models with good skills in predicting flood extent areas, water levels and velocities (Di Baldassarre *et al.*, 2009c). Unfortunately, the choice of a particular model is still often done a priori, without considering the aforementioned principles and criteria. The next section provides useful information to assist modellers in selecting appropriate tools for modelling inundation processes.

5.2.2 Model comparison

In recent years, many studies comparing different approaches for floodplain modelling have been published. It should be noted that the outcomes of this type of studies are often associated with the specific test sites. This section summarizes the outcomes of two comparison studies performed by Horritt and Bates (2002), in a rural river reach, and Hunter *et al.* (2008), in an urbanized area.

Horritt and Bates (2002) evaluated HEC-RAS, LISFLOOD-FP and TELEMAC-2D in a 60-km reach of the River Severn (UK) using independent calibration data from both hydrometric and SAR sources. They concluded that all three models were capable of predicting flood extent and travel times to similar levels of accuracy at optimum calibration. However, it should be noted that flow on this river reach is confined to a relatively narrow valley, and one might expect more complex inundation patterns in wider floodplain areas, and in that case the 2D approach may prove more effective than 1D (Horritt and Bates, 2002). Differences between the models emerged according to the type of calibration data used when the models were used in predictive mode (Chapter 6). In this case, both HEC-RAS and TELEMAC-2D were capable of making equally good predictions of inundated area, whether calibrated against flood wave travel times or against inundated area data from another event. By contrast, predictions of flood extent from LISFLOOD-FP were found to be significantly poorer when the model is calibrated against flood wave travel time. The outcome indicates that, when models are used in predictive mode, one should be careful in using them to predict something different from what was used for their calibration, especially when more conceptual models (e.g. LISFLOOD-FP) are utilized.

Hunter *et al.* (2008) tested and compared six 2D models (DIVAST, DIVASTTVD, TUFLOW, JFLOW, TRENT and LISFLOOD-FP) in terms of their ability to simulate inundation patterns in an urban catchment within the city of Glasgow

(Scotland, UK). This comparison pointed out that all the models produce plausible results in simulating urban flood inundation processes. Hunter *et al.* (2008) also pointed out that modern LiDAR DTMs (Chapter 4) are sufficiently accurate and resolved for simulating floods in urban areas. Moreover, they found that flows in urban environments are characterized by numerous transitions to supercritical flow and numerical shocks. However, the effects of these tend to be localized and therefore do not seem to significantly affect the flood propagation and inundation processes overall.

5.3 MODEL IMPLEMENTATION

In addition to the selection of a numerical tool for modelling flood propagation and inundation processes, modellers have to choose a way to discretize time, schematize the geometry and parameterize the roughness of rivers and floodplains.

5.3.1 Time discretization

The time derivative term of the SWE (Chapter 3) can be discretized in several ways, using either explicit or implicit schemes. The consequences of selecting one method or the other are in the complexity of the algorithms required to solve the resulting equations and in the stability of the numerical model (Hunter *et al.*, 2007).

Explicit solutions are often favoured as they are simple to code and allow straightforward integration of models within a dynamic GIS environment (Burrough, 1998). Also, the advantage of explicit solutions is the ease of implementation and minimal code changes required to run simulations in parallel (Neal *et al.*, 2009b). However, the model time step, Δt, must be selected very carefully by the modeller. As explicit numerical schemes are conditionally stable, Δt must be small enough to satisfy the Courant condition and prevent instabilities developing in the numerical solution (Hunter *et al.*, 2007). This might lead to computational time steps that are very small (potentially of the order of seconds) compared to the physical phenomena under consideration (Cunge *et al.*, 1980).

By contrast, dependent variables in an implicit scheme are additionally evaluated in terms of unknown quantities at the new time step, $t + \Delta t$, and use either a matrix or iterative technique to obtain a solution. These methods couple together all cells within the solution domain, which allows flow behaviour to be transmitted through the entire model grid. The price for this communication between distantly located cells is increased code complexity and computational cost (Hunter *et al.*, 2007). Nevertheless, implicit schemes ensure the unconditional stability of the solution and allow larger time steps (potentially of the order of hours) more

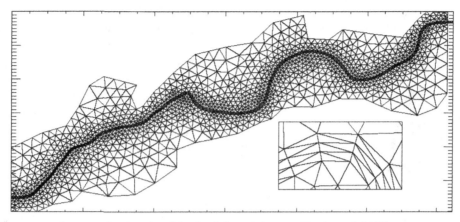

Figure 5.4 Example of unstructured computational mesh and details of channel elements (Horritt *et al*., 2007).

compatible with the slow evolution of flood events. More details on time discretization can be found in Cunge *et al*. (1980).

5.3.2 Geometry

The representation of the model geometry depends on the type of numerical tool. For instance, 2D models typically require DEMs, while 1D models use a series of cross sections (Table 5.2). Moreover, while finite difference models typically need structural computational grids, finite element (or finite volume) models use unstructured meshes.

The schematization of the geometry in 2D finite difference models is directly based on the DTM. It has been demonstrated that the spatial resolution has a striking effect on hydraulic predictions (Hardy *et al*., 1999). Horritt and Bates (2001) evaluated the effects of grid size on flood inundation modelling and showed that the maximum performance is reached at a resolution between 50 m and 100 m, after which no improvement was found with increasing resolution, provided that results are re-projected in a high-resolution DTM. However, it is worth recalling here that the value of DTMs in flood inundation modelling depends not only on their resolution, but also on their accuracy (Chapter 4). Also, it is worth mentioning that these indications are valid in rural floodplains (see Chapter 8 for urban areas).

As mentioned above, 2D finite element (or volume) models typically use unstructured meshes. The advantage of using unstructured mesh is that the number of computational nodes needed for the description of topographical discontinuities that influences the inundation process, such as road embankments, is minimized. Figure 5.4 shows an example of computation mesh for finite element (and volume) models generated using the CheesyMesh algorithm (Horritt, 2000; Horritt *et al*., 2007). By analysing the inset of Figure 5.4 one can observe that the main channel is represented by elements elongated in the direction of flow, with three nodes across the channel. This enables the trapezoidal form of the channel and cross-channel velocity profiles to be represented with a minimum

number of elements (Horritt *et al*., 2007). By contrast, unstructured meshing is used on the floodplain, with a smooth transition between the channel and maximum element sizes (Figure 5.4). Typically, floodplain topography is sampled onto each node from the DEM using nearest neighbour sampling, while channel node elevations are sampled from the ground-surveyed cross sections, and interpolated along the channel (Horritt *et al*., 2007).

Despite many studies investigating the impact of mesh resolution on hydraulic modelling (e.g. Hardy *et al*., 1999; Yu and Lane, 2006; Horritt *et al*., 2007), drawing comprehensive guidelines is still not possible. However, a good strategy is to follow the recommendations of Hardy *et al*. (1999). They concluded that at the beginning of new modelling exercises, prior to more complex calibration processes, one should construct at least four meshes of different spatial resolutions to determine the envelope of response to spatial resolution.

In 1D modelling, the geometry of rivers and floodplains is typically described by cross sections, which can be either derived via traditional ground surveys or extracted from high-resolution DTMs. Guidelines for optimal cross-section distances are provided in Chapter 4. For a correct schematization of 1D models, cross-section lines should be approximately perpendicular to the direction of flow (Castellarin *et al*., 2009). However, given that the average direction of the flow may vary significantly, a model designed for flood flow computation may not reproduce the low-flow dynamics and vice versa (Samuels, 1990). Moreover, the local direction of flow is strongly dependent on the riverbed morphology, which, in turn, influences the layout of cross-section lines. Practically, there are three different ways to conduct the topographical survey of a riverbed: (i) cross-section lines are approximately normal to the direction of flow in the main channel; (ii) cross-section lines are approximately normal to the average direction of flow within the whole flood envelope; (iii) cross-section lines are approximately normal to the direction of flow in the main channel and in the floodplain areas (Figure 5.5; Castellarin *et al*., 2009).

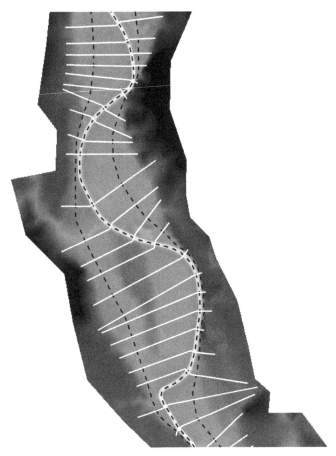

Figure 5.5 Example of cross-section lines (white lines) approximately perpendicular to both floodplain and in-channel flow directions (dashed lines).

The first layout (case i) is rather frequent in practice. Its main drawback consists of overestimating the cross-sectional area in floodplains and therefore it generally underestimates water levels. The second layout (case ii) is not capable of correctly reproducing the real extension of the main channel under low-flow conditions. The third layout (case iii; Figure 5.5) represents an acceptable trade-off when the modeller tries to get a satisfactory reproduction of the hydraulic behaviour of a reach for a wide range of flow conditions, from low flows to flood flows. The three layouts do not differ strongly from one another when the main channel meandering is limited. However, in some cases, the significant differences among these three layouts may impact the results of the 1D hydraulic computations (Castellarin *et al.*, 2009).

5.3.3 Roughness

Roughness coefficients may arguably be estimated in the field with a high degree of precision (Chow, 1959; Cunge, 2003). However, it has not been proved that such physically based parameters are capable of providing accurate predictions from single model

realizations (e.g. Beven, 1989; Bates *et al.*, 2005; Di Baldassarre *et al.*, 2010b). This is because roughness coefficients are required to represent a range of different sources of energy loss, whose explicit treatment within a particular model varies with code dimensionality and process representation decisions (Lane and Hardy, 2002; Hunter *et al.*, 2007). Also, the predetermination of model parameters at each computational grid point is rarely possible due to experimental constraints (Beven, 2006) and scaling problems, such as differences between the measurement scale, model grid scale, and the scale at which the basic algorithmic process descriptions are derived (Hunter *et al.*, 2007). Moreover, it should be noted that the resistance to flow is theoretically a function of water depth. Thus, while the parameterization of the model geometry is usually based on a measurement approach (using DEM and cross sections, see above), the parameterization of the flow resistance (i.e. identification of the roughness coefficients) is usually performed using a calibration approach (Horritt, 2005).

The roughness coefficients can be specified individually for each computational node (Hunter *et al.*, 2007). However, given that the roughness coefficients are often identified via calibration, such an approach would lead to large uncertainty in the estimation of the roughness parameters (principle of parsimony, Figure 5.2). Hence, the domain of flood inundation models is typically divided into a few classes (often only two: floodplain and river) where lumped values of the roughness coefficients are used (Horritt *et al.*, 2007). The calibration of the roughness coefficients is discussed in Chapter 6.

5.4 CONCLUSIONS AND PERSPECTIVES

One of the most important questions to address in building hydraulic modelling is 'how simple can a model be and still be physically realistic?' (Hunter *et al.*, 2007). As early as 1975, Price demonstrated that simplified flood propagation models usually meet the requirements of practical applications. Since then, the scientific literature has presented a large number of inundation models of reduced complexity (Hunter *et al.*, 2007).

The selection criteria presented in this chapter indicate that a model should be parsimonious, computationally efficient, numerically stable, and able to capture the dominant processes and predict the desired quantities (e.g. flood extent, flood water levels). In addition, model selection should also be based on the actual availability and quality of data as well as the purpose of the model. For instance, while real time forecasting typically does require fast models (Szöllösi-Nagy and Mekis, 1988), in flood risk mapping the computational time is not as crucial as the skill of the inundation models in predicting water levels, velocities, and flood extent areas (Di Baldassarre *et al.*, 2009c).

Given the dependence on the particular application and the fact that 'all models are wrong, but some are useful' (Box, 1976), conclusive rules could not be provided. However, to support the selection of the most appropriate model, this chapter provided a concise description of numerical tools that are commonly used in flood risk mapping and an indication of their potential application, the necessary input data, the type of outputs, and the order of magnitude of the computation time.

Lastly, the chapter discussed the schematization of the geometry of the model domain and the parameterization of the flow resistance. The chapter pointed out that, while the former is usually carried out with a measurement approach (use of DEM or cross sections), the latter is typically performed using a calibration, as roughness coefficients are difficult (if not impossible) to measure. Thus, calibrated roughness parameters should be recognized as being effective values that may not reproduce (but do represent) subgrid heterogeneities and may not have a physical interpretation outside of the model within which they were calibrated (Beven, 2000; Hunter et al., 2007).

In terms of perspectives, although many flood inundation models have been proved to be valuable in supporting floodplain management, understanding sediment dynamics and flood risk mitigation (Horritt et al., 2007), there is still a need to improve our ability to observe, analyse and model floodplain systems. First, many current methods are based on theories and models that were developed for specific case studies, mainly in temperate regions. As a consequence, they might neglect some processes that could be relevant for many other floodplain systems, such as direct precipitation, evapotranspiration, groundwater interaction, or sediment transport (Bates et al., 2000). For instance, most modelling tools might not be suitable for tropical and subtropical floodplains, where infiltration and evaporation often play an important role in the inundation processes. Also, several methods suffer in dealing with different time and space scales, as processes dominant at the smaller scales can be of less importance at larger scales (Bloeschl, 2006; Fenicia et al., 2008). Second, traditional approaches tend to focus on natural processes only, while humans are considered external to the floodplain system. However, given the growing impact of human activities, natural (pristine) floodplains have become more and more uncommon (Sanderson et al., 2002). Thus, there is a need for better understanding of the coupled human–ecosystems. Third, current floodplain models are often based on the assumption of stationarity and their predictions are usually tested by simply reproducing the past (Wagener et al., 2010). However, traditional calibration and validation based on past data, though necessary in many instances, is insufficient in many cases because of environmental changes (Di Baldassarre et al., 2011b). Lastly, there is still a general tendency to neglect (or not explicitly estimate and therefore attempt to reduce) all the relevant sources of uncertainty intrinsic to any floodplain modelling exercise, such as inaccurate input data, imperfect model structure, or inadequate model parameterization (see Chapter 7).

In the near future, current research work might lead to flood inundation models that have reduced predictive uncertainty (Bates et al., 2005). For instance, we might improve our ability to represent flow–vegetation interactions and flows in the near-wall region (Wilson et al., 2005). In this context, an improved understanding of the generation of friction and turbulence by vegetation combined with enhanced retrieval of plant biomechanical properties from remotely sensed data has great potential (Woodhead et al., 2009). Furthermore, the specification of appropriate schemes that are capable of handling dynamic wetting and drying effects over complex, low-lying topography, is still poorly understood (Bates and Horritt, 2005), although much research has already been undertaken in this area (Lynch and Gray, 1980; Defina, 2000; Bates et al., 2005).

5.5 EXERCISES

5.1. Make a practical example of the principle of parsimony.

5.2. Illustrate the types of numerical tools commonly used in flood propagation and inundation modelling and indicate their potential applications.

5.3. Explain why the finite element approach is able to describe relevant topographical features while minimizing the number of computational nodes.

5.4. Provide a qualitative description of the potential impact on hydraulic modelling of the different approaches to extract cross sections from a DTM.

5.5. Explain why the parameterization of the flow resistance, i.e. identification of roughness coefficients, is typically performed using a calibration approach and not a measurement approach.

6   Model evaluation

This chapter discusses the evaluation of flood inundation models. After the introduction of basic concepts, the chapter presents performance measures that are commonly used to compare model results and observations. The calibration and validation of hydraulic models is also discussed. Lastly, the chapter introduces methodologies, recently proposed in the scientific literature, which can be used to cope with uncertainty in hydraulic modelling.

6.1   CONCEPTS

A rigorous discussion of model evaluation requires a clear distinction between code verification and model validation. To this end, this section refers to definitions reported in Refsgaard (2001) and Hunter *et al.* (2007).

6.1.1   Code verification

A 'model code' is a generic computer program that can be used for different river reaches and catchments without modifying the source code. It summarizes the modeller's perception of how a river system works under different flow conditions in a specific mathematical formulation. A model code can be verified. Code verification involves comparison of the numerical solution generated by the code with one or more analytical solutions or with other numerical solutions. Verification ensures that the computer program accurately solves the equations that constitute the mathematical model. In recent years, analytical solutions of the SWE have provided useful tests for a variety of model codes as they have allowed the assessment of hydrodynamic schemes without introducing additional sources of uncertainty, such as topography, boundary conditions and roughness coefficients. An obvious practical recommendation when using numerical codes is to check how the model code was verified. More details and references about code verification can be found in Hunter *et al.* (2007).

6.1.2   Model validation

A 'model' is a site application of a code to a particular river reach, including specific input data and parameter values. Model validation is defined as the process of assessing whether a given site-specific model is capable of making accurate predictions, defined with respect to the specific application, for periods that are outside the calibration period. A model is said to be validated if its accuracy and predictive capability in the validation period is proved to lie within acceptable limits for a particular practical purpose (Refsgaard, 2001; Hunter *et al.*, 2007).

6.2   PERFORMANCE MEASURES

The assessment of flood inundation models for a specific case study requires comparison of model results with observations. This section reports some performance measures that are commonly utilized to quantify the error (or the match) between models and observations.

6.2.1   At-a-point time series (hydrographs)

Traditionally, flood inundation models have been calibrated and validated using at-a-point time series, such as stage (or flow) hydrographs recorded in gauging stations. Given the relevant uncertainty of stage–discharge rating curves (Chapter 4), recorded water levels are often more valuable than observed discharges. Thus, to evaluate hydraulic models, internal stage hydrographs should be preferred to flow hydrographs. Figure 6.1 shows an example of observed and simulated stage hydrographs.

There are several methods to compare observed and simulated time series (Beven and Freer, 2001) and they all unavoidably have a bias towards one specific characteristic of the hydrograph (Hunter *et al.*, 2005a). The Nash and Sutcliffe (1970) efficiency

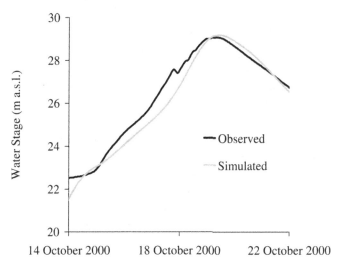

Figure 6.1 Example of observed (black) and simulated (grey) stage hydrographs.

(*NSE*) is one of the most commonly used measures (Cameron *et al.*, 1999):

$$NSE = 1 - \frac{\sum\limits_{t=1}^{T} [z_s(t) - z_o(t)]^2}{\sum\limits_{t=1}^{T} [z_m - z_o(t)]^2} \quad (6.1)$$

where $z_o(t)$ and $z_s(t)$ are observed and simulated water levels at time t, z_m is the mean value of the observations, and T the number of time steps. The *NSE* ranges between $-\infty$ and $+1$ with all models scoring below zero being no better than using the mean of observations. Given that *NSE* is based on the sum of error variance (equation 6.1), it is sensitive to differences in both maximum values and timing of flood peak (Hunter *et al.*, 2005a). The use of error variance as the performance measure is most suitable when errors are of mean zero, normally distributed with constant variance, and not correlated (Beven, 2001). However, it should be noted that hydrometric data may often violate this assumption (Montanari, 2005). Hence, alternative measures able to account for the presence of either autocorrelation or heteroscedasticity have been proposed (e.g. heteroscedastic maximum likelihood estimator; Sorooshian *et al.*, 1983).

6.2.2 Spatially distributed, continuous point data (high water marks)

At-a-point time series, such as stage hydrographs, do not test the distributed model performance, which is often the type of prediction that hydraulic models are expected to deliver. To this end, high water marks (i.e. post-event measured maximum water levels; Chapter 4) are a valuable type of evaluation data. High water marks are very useful, in particular, when the main purpose of hydraulic modelling is the simulation of flood profiles for the

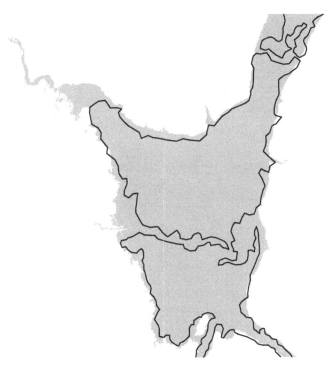

Figure 6.2 Example of simulated flood extent area (in grey) and contour of a SAR-derived flood extent map (black line).

design of levee systems. Obviously, one should consider that this type of evaluation data does not provide any information about flood propagation (e.g. wave speed). Flood water levels derived from SAR data (Section 4.2.3) are another example of spatially distributed, continuous point data that can be used for model evaluation.

The comparison of model results with high water marks or SAR-derived flood water levels can be done using the mean absolute error (*MAE*):

$$MAE = \frac{\sum\limits_{x=1}^{N} |Z_s(x) - Z_o(x)|}{N} \quad (6.2)$$

where $Z_s(x)$ and $Z_o(x)$ are the simulated and observed water levels at the river chainage x, and N is the number of points where high water marks are available.

6.2.3 Spatially distributed, binary pattern data (flood extent maps)

As mentioned in Chapter 4, binary (wet/dry) maps of flood extent derived from SAR images are widely used to evaluate flood inundation models and compare different models (Wright *et al.*, 2008). Figure 6.2 shows an example of a flood extent map derived from ERS-2 SAR in the Lower Dee (UK) during the 2006 inundation, in addition to the results of a TELEMAC-2D model (Di Baldassarre *et al.*, 2010b). Simulated flood extent maps are typically derived by re-projecting the simulated water levels onto the LiDAR DTM;

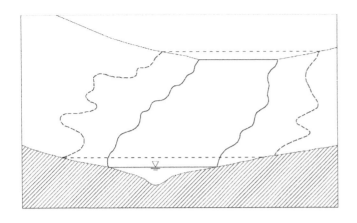

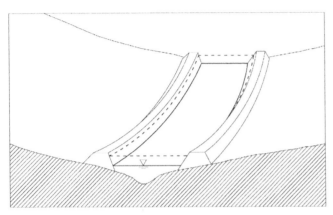

Figure 6.3 Model evaluation using flood extent maps. Top row: example of undefended floodplain where small changes in water levels produce large changes in lateral flood extent. Bottom row: flood shoreline is constrained by embankments; in this case flood extent maps are not useful evaluation data. (Sketch kindly drawn and provided by Domenico Di Baldassarre.)

thresholds around 10 cm of water depth are used to differentiate wet and dry areas.

It should be noted that flood extent maps should be used with caution in assessing model performance. In fact, the evaluation of flood inundation models using observed binary maps, such as SAR-derived flood extent maps, requires the assumption that the correct simulation of inundation extent necessarily implies an accurate reproduction of water depths across the floodplain. While this assumption is potentially reasonable for undefended or unrestricted rural reaches (Figure 6.3, top panel), this is not always the case. For instance, during high-magnitude events where the valley is entirely inundated or when the flood shoreline is constrained by slopes or defences (e.g. levees, embankments), large changes in water levels may produce only small changes in lateral flood extent (Figure 6.3, bottom panel). In this second case, flood extent maps are not useful for model assessment (Hunter *et al.*, 2005a) and high water marks are more useful to evaluate model performance.

Table 6.1 *Contingency table*

	Observed wet	Observed dry
Simulated wet	A	B
Simulated dry	C	D

The literature provides many performance measures to compare SAR-derived and simulated (binary) flood extent maps. The comparison is typically based on a contingency table, which reports the number of pixels correctly predicted as wet or dry, and under-prediction and over-prediction (Table 6.1).

Two measures of fit commonly used in flood inundation modelling are

$$F_1 = \frac{A}{A + B + C} \qquad (6.3)$$

$$F_2 = \frac{A - B}{A + B + C} \qquad (6.4)$$

where A is the size of the wet area correctly simulated by the model, B is the area simulated as wet that is observed dry (over-prediction), and C is the area observed as wet that is not simulated by the model (under-prediction). F_1 ranges from 0 to 1, while F_2 ranges between -1 and 1. These measures of fit were found to be valuable in flood inundation modelling (Horritt *et al.*, 2007). In F_2, the term $-B$ in the numerator of equation (6.4) is used to penalize model over-prediction. This may be appropriate as the flood extent area simulated by the models tends to be underestimated compared to the SAR-derived flood extent area. In fact, SAR images of floods give an aggregated response of all flood-related processes (floodplain inundation as a result of excess rainfall, bank overtopping, backwater effects, complex 2D and 3D flows, etc.), whereas models only reproduce a few of these to inundate the floodplain (Schumann *et al.*, 2009b). Other performance measures commonly used to compare simulated and observed flood extent maps are (A, B, C and D as defined in Table 6.1)

$$Bias = \frac{A + B}{A + C} \qquad (6.5)$$

$$PC = \frac{A + D}{A + B + C + D} \qquad (6.6)$$

The *bias* can be useful to summarize aggregate model performance, i.e. under-prediction or over-prediction. Although widely used, *PC* (predicted correct) is not recommended, as the values for D (Table 6.1) are usually larger than the other categories and may also be trivially easy to predict. Thus, in many instances, *PC* will tend to provide an overly optimistic assessment of model performance (Aronica *et al.*, 2002; Schumann *et al.*, 2009b).

As a final remark, it is worth noting that all these aggregate measures, (6.3)–(6.6), do not allow the analysis of the variability of model performance in space. However, given that uniform model parameterization is typically used to ensure a computationally tractable problem (see below) the use of a lumped performance measure can be considered appropriate (Aronica *et al.*, 2002). Also, it should be mentioned that all these performance measures give the same weight to all the pixels of the model domain. In this context, Pappenberger *et al.* (2007) argued that model performance should be assessed using a vulnerability weighted approach, where more weight is given to correct predictions of inundation patterns in the proximity of critical infrastructures, such as hospitals, and less to rural areas, such as empty pastures.

6.3 CALIBRATION AND VALIDATION

Model calibration is undertaken to identify appropriate values for parameters such that the model can reproduce observed data. In flood inundation modelling, the roughness coefficients assigned to the main channel and floodplain are used as calibration parameters (Woodhead *et al.*, 2009). Some form of calibration is always required to apply a hydraulic model to a particular river reach for a given flood event (Bates *et al.*, 2005). As mentioned, some hydrologists argue that roughness coefficients should be assessed using engineering judgment, and physically implausible roughness values should be used as the evidence that the model does not reproduce reality (Cunge, 2003). This statement may be correct under the assumption that all data are error free and the model structure is perfect, and that the point roughness values derived by observation can adequately reflect the spatial variability in momentum losses on heterogeneous floodplains that affect effective roughness values at the model discretization scale. Given that this is never the case, reliable flood inundation models need to accurately predict water levels and flood extent (rather than roughness) and therefore calibration (or sensitivity analysis) is recommended (Di Baldassarre *et al.*, 2010b). As such, parameter values calculated by the calibration of models should be recognized as being effective values that may not have a physical interpretation outside of the model structure within which they were calibrated (Bates *et al.*, 2005). For instance, Horritt *et al.* (2007) demonstrated that roughness coefficients, as averaged in space and time within a particular numerical model, are not fixed quantities representing an identical set of physical processes. By contrast, the roughness coefficients are scale-dependent effective parameters, which represent all those energy losses not represented explicitly by the model physics at the space and time resolution selected by the modeller. Even when simulations are conducted with models that have a very similar physical basis and are applied to the same model grid, significantly different optimum calibrated parameter values are obtained. Friction values thus vary not only with model space and time resolution, and with the physical processes included within the model, but also with the particular implementation of these processes. Referring

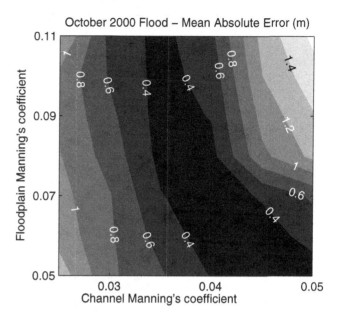

Figure 6.4 Model calibration: contour plot of *MAE* (m) between the maximum simulated water levels and high water marks measured after the October 2000 flood.

to the friction loss parameters in a numerical hydraulic model as 'Manning's *n*' is no more than a semantic convenience, and does not truly indicate a shared conceptual basis with the point scale values (Horritt *et al.*, 2007). For such effective parameters it therefore becomes more difficult to specify a priori an appropriate distribution within which to search for a performance optimum, meaning that published tables of Manning's coefficients (Chow, 1959), or more up to date approaches for 1D models (Fisher and Dawson, 2003), should be regarded, at best, as only a guide to the likely range (Horritt *et al.*, 2007).

As an example, Figure 6.4 shows the calibration of a HEC-RAS model of a 98-km reach of the River Po (Italy) between Cremona and Borgoforte (Figure 6.5). The calibration was carried out by comparing the simulated maximum water elevations with the high water marks surveyed in the aftermath of the October 2000 flood. The use of high water marks, instead of flood extent maps, is justified by the high magnitude of the October 2000 event and the presence of major embankments that constrain the flood extent (Figure 6.3).

Figure 6.4 shows the model response in terms of *MAE* (m). The best fit model is characterized by Manning's coefficients of around $0.04 \text{ m}^{-1/3} \text{ s}$ for the channel and $0.09 \text{ m}^{-1/3} \text{ s}$ for the floodplain. However, by analysing Figure 6.4, one can observe that there are different parameter sets that give *MAE* below 0.40 m, which is a good performance, as 0.40 m is the expected accuracy of high water marks (Chapter 4). Furthermore, it is interesting to note that these optimal parameter sets lie inside a certain (hyperbolic shape) area (Figure 6.4). This is an example of parameter compensation: the decreasing Manning's floodplain coefficient

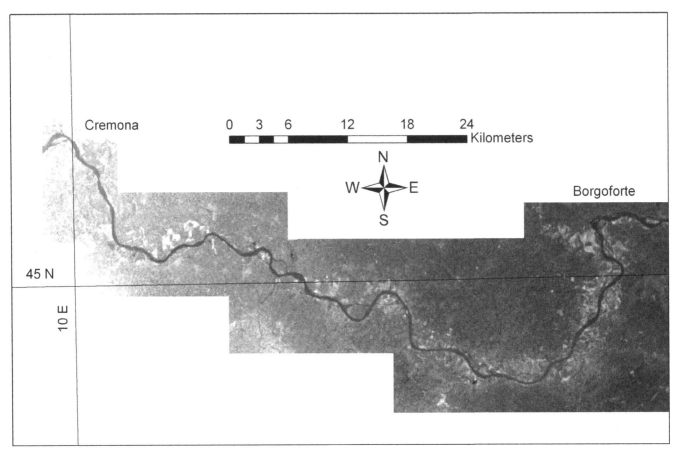

Figure 6.5 River Po (Italy) between Cremona and Borgoforte (SRTM topography).

is compensated by an increasing Manning's channel coefficient. Figure 6.4 reflects the frequent situation when different combinations of effective parameter values may fit calibration data equally well. Such equifinality in flood inundation modelling has been well documented (Romanowicz *et al.*, 1996; Aronica *et al.*, 1998; Hankin *et al.*, 2001; Romanowicz and Beven, 2003; Bates *et al.*, 2005) and uncertainty analysis techniques have been developed and applied in response (see below).

As mentioned, model validation is the process of evaluating if a given site-specific model is capable of making accurate predictions, defined with respect to the specific application, for periods outside a calibration period. As an example of model validation, Figure 6.6 shows the validation of the aforementioned HEC-RAS model of the River Po between Cremona and Borgoforte (Figure 6.5). The validation was performed by referring to the June 2008 flood. Given the low magnitude of this event (Di Baldassarre *et al.*, 2009a), it was found that small errors in water levels imply large errors in inundation extent. Thus, SAR-derived inundation widths are valuable validation data. Specifically, the validation was performed by using the inundation widths derived from satellite imagery of the June 2008 flood (ENVISAT-ASAR

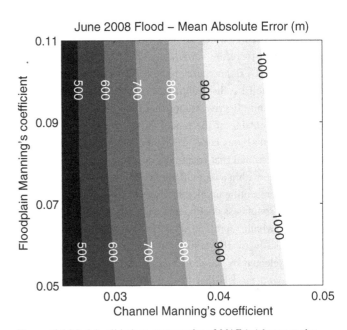

Figure 6.6 Model validation: contour plot of *MAE* (m) between the maximum simulated inundation widths and using the SAR-derived flood extent during the June 2008 flood.

in WSM). The results of the validation exercise are shown in Figure 6.6.

By comparing Figure 6.4 and Figure 6.6, one can observe that the overall model response is rather different. In fact, it has been widely shown that effective roughness coefficients may be different when evaluated for flood events of different magnitude (e.g. Romanowicz et al., 1996; Aronica et al., 1998; Horritt and Bates, 2002; Horritt et al., 2007). More specifically, the comparison between Figures 6.4 and 6.6 shows that the calibrated model (Manning's coefficients of around 0.04 $m^{-1/3}$ s for the channel and 0.09 $m^{-1/3}$ s for the floodplain; Figure 6.4) fails to reproduce the low-magnitude flood event of June 2008, as its *MAE* in simulating the June 2008 inundation is equal to around 900 m (Figure 6.6). This confirms several studies on the evaluation of flood inundation models: a well-calibrated flood inundation model may perform poorly when it is used to predict events of different magnitude (Di Baldassarre et al., 2009a).

6.4 UNCERTAINTY ANALYSIS

6.4.1 Uncertainty in flood inundation modelling

The scientific literature has widely shown that flood inundation modelling is affected by relevant uncertainty (Aronica et al., 1998; Bates et al., 2005; Hunter et al., 2005a, 2007; Pappenberger et al., 2005; Di Baldassarre et al., 2009a, b, c, 2010b). Yet, it is largely recognized that there is a need for better understanding and identification of the sources of uncertainty as well as quantifying the model uncertainty (Solomatine and Shrestha, 2008).

In hydraulic modelling, uncertainty is caused by different sources (e.g. Gupta et al., 2005): input data (e.g. boundary conditions), calibration data (e.g. high water marks, flood extent maps), model parameters (e.g. roughness coefficients), and imperfect model structure. Some authors classify these sources by differentiating between epistemic and aleatory uncertainty (Beven, 2008). The epistemic uncertainty refers to the state of knowledge of a physical system and our ability to measure and model, while aleatory uncertainty represents the randomness and variability (both in space and time) observed in nature.

In recent years, a number of methods have been proposed to estimate uncertainty. They can be classified as follows (e.g. Montanari, 2007; Solomatine and Shrestha, 2008): (i) analytical methods, (ii) approximation methods, (iii) simulation and sampling-based methods, (iv) Bayesian methods, (v) methods based on the analysis of model errors, and (vi) methods based on fuzzy set theory. More details about different methods for uncertainty estimation are reported in Gupta et al. (2005). What should be noted here is that, given that any uncertainty assessment method is conditioned on some assumptions, the choice of the most appropriate technique should be made on the basis of the knowledge of the river/floodplain system, the input and output data, and the available modelling approaches. In general, different techniques for uncertainty estimation may be valuable, depending on the scope of the analysis. Anyhow, it is extremely important that all of the underlying assumptions are stated explicitly and the consequent limitations discussed in full detail (Montanari, 2005), regardless of which particular technique is applied.

6.4.2 The GLUE framework

This section provides a brief description of GLUE (generalized likelihood uncertainty estimation; Beven and Binley, 1992), which is an informal Bayesian approach to estimate uncertainty. GLUE has been widely used for uncertainty estimation in flood inundation modelling (Romanowicz et al., 1996; Aronica et al., 1998; Romanowicz and Beven, 2003; Bates et al., 2005; Hunter et al., 2005a, 2007; Pappenberger et al., 2005; Horritt et al., 2007; Di Baldassarre et al., 2009b, 2010b) as it is easy to apply and can account for all sources of uncertainty in hydraulic modelling, either explicitly or implicitly (Montanari, 2005). Also, GLUE does not need strong assumptions about the nature of the statistical properties of the residuals. However, GLUE requires a number of subjective decisions (see below) and, therefore, has been criticized by part of the scientific community (e.g. Mantovan and Todini, 2006). Nevertheless, Vrugt et al. (2009) showed that formal Bayesian approaches can generate estimates of total predictive uncertainty very similar to those of informal Bayesian approaches, such as GLUE.

GLUE rejects the concept of a single optimum model and assumes that prior to input of data into a model, all models have an equal likelihood of being acceptable (Montanari, 2005). The GLUE framework requires the following steps (Freer et al., 1996):

(1) Identification of the parameters that most affect the model output. Then, a high number of parameter sets are generated via uniform sampling, or incorporating prior knowledge about the distribution of parameters (Montanari, 2005).

(2) All the models are run and the outputs compared to observed data. The performance of each model is assessed through goodness-of-fit measures (see Section 6.2).

(3) Performance evaluation includes rejecting some models as non-behavioural. All models and their corresponding parameter sets that provide a likelihood measure that reaches a minimum threshold are retained as behavioural models. This step has a clear subjective nature. Thus, this decision has to be transparent and unambiguous (Montanari, 2005).

(4) Let us suppose that the hydraulic model is evaluated using high water marks, i.e. observed maximum water levels, $Z_o(x)$. Thus, the maximum water level in x is simulated by n

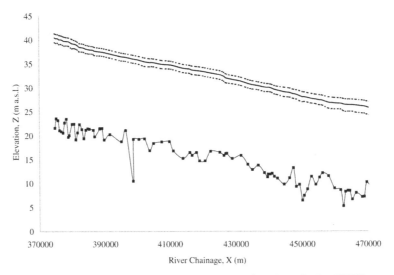

Figure 6.7 Ensemble simulation of the 2000 flood event: median and uncertainty bounds estimated using GLUE.

behavioural models, which provide the simulations $Z_{s,i}(x)$, $\{i = 1, 2, \ldots, n\}$, each one characterized by its own likelihood measure.

(5) The calculated likelihoods are rescaled to produce a cumulative sum of 1. Hence, one associates to each $Z_{s,i}(x)$ its rescaled likelihood weight.

(6) The $Z_{s,i}(x)$ are ranked in ascending order and a probability of not exceedance is assigned to each of them, which is equal to the cumulated sum of the rescaled likelihood weights up to the considered $Z_{s,i}(x)$.

(7) A cumulative distribution function of simulated water levels is built, with the highest $Z_{s,i}(x)$ associated to a probability of not exceedance equal to 1. This allows uncertainty bounds corresponding to an assigned confidence level to be derived, in addition to a median simulation (Montanari, 2005).

6.4.3 Uncertainty estimation

An example of uncertainty estimation by applying the GLUE method is reported here. The application refers to the HEC-RAS model of the River Po (Italy) between Cremona and Borgoforte (Figure 6.5).

As mentioned, the first step requires the selection of a subset of behavioural models (Beven, 2006) satisfying a threshold criterion. For this exercise, the high water marks of the 2000 flood were used as calibration data (Figure 6.4), and all the couples of Manning's channel and floodplain coefficients giving a mean absolute error higher than 1 m (Figure 6.4) were rejected, while the others were considered as behavioural models. As mentioned, this step has a clear subjective nature and has to be transparent and unambiguous (Montanari, 2005). However, it should be noted that transparency in the decision-making does not eliminate this

subjectivity (Hunter *et al.*, 2005b). Thus, only the models satisfying this threshold condition were used to make an ensemble simulation of the 2000 flood event. More specifically, following the GLUE framework, each behavioural simulation, i, was associated to a likelihood weight, W_i, ranging from 0 to 1. The weight was expressed as a function of the measure of fit, MAE_i, of the behavioural models:

$$W_i = \frac{\max(MAE_i) - MAE_i}{\max(MAE_i) - \min(MAE_i)} \quad (6.7)$$

where $\max(MAE_i)$ and $\min(MAE_i)$ are the maximum and minimum value of the MAE of the ith behavioural model. Then, the likelihood weights were rescaled to a cumulative sum of 1 and the weighted percentiles (5th, 50th and 95th), representing the likelihood weighted uncertainty bounds, were computed (Figure 6.7). A Matlab code to compute weighted percentiles is included in the password-protected electronic resources of this book (*Folder: Matlab*).

By considering different parameter sets, this example explicitly evaluated the effects of parameter uncertainty (Figure 6.7). Similarly, if one considered either different model input and output or different model structures, the effects of either observation uncertainty or model structural uncertainty, respectively, would be explicitly assessed. However, it should be noted that there is an interaction among the different sources of uncertainty. For instance, parameter uncertainty is strictly related to observation uncertainty because imperfect input data may induce identifiability problems, and equifinality in the estimation of model parameters. Thus, one should not expect to always be able to treat each source of uncertainty individually, but, rather, to implicitly deal with different sources of uncertainty (Montanari, 2005).

6.5 CONCLUSIONS AND PERSPECTIVES

This chapter dealt with the assessment of hydraulic models. After the distinction between code verification and model validation, it presented a number of goodness-of-fit measures that are commonly used in flood inundation modelling. The chapter also discussed the estimation of effective parameter values through calibration and showed that this process is affected by a number of error sources that cast some doubt on the certainty of calibrated parameters (Aronica et al., 1998; Bates et al., 2005). Essentially, these errors relate to numerical approximations associated with the discrete solution of the controlling flow equations; inaccuracies of calibration data, and the fact that input data might be inadequate to represent heterogeneous river reaches. Hydraulic models therefore always require the estimation of effective parameters that partly compensate for these sources of error (Romanowicz and Beven, 2003). Lastly, the chapter provided a general overview of the techniques that can be used to cope with uncertainty in hydraulic modelling and a description of the most applied methodology, namely GLUE.

Uncertainty estimation in hydraulic models still requires additional efforts. In fact, although researchers and modellers are well aware that a significant approximation affects the output of flood inundation models, environmental agencies, river basin authorities and engineering consultancies hardly ever apply recent advances in uncertainty analysis and probabilistic flood mapping. This may be caused by the difficulties in transferring relative know-how between scientists and end-users as well as the complexity of uncertainty estimation methods, which may have hindered their practical application. To facilitate a wider application of these methods, the development of clear and mature guidance on methods and applications is still needed.

6.6 EXERCISES

The exercises of this chapter require the use of HEC-RAS (see Chapter 5). To this end, the electronic resources of this book include the HEC-RAS software (Folder: Hec-Ras\Installation-files). Installation instructions and step by step indications on how to import data, how to run simulations and how to visualize the results are reported in the user manuals, which are also included in electronic resources of this book. The interested reader might also check on the website of the US Army Corps of Engineers for possible updated versions of the software.

Hydraulic modelling

In these exercises, HEC-RAS is used to simulate the hydraulic behaviour of the 98-km reach of the River Po (Italy), between Cremona and Borgoforte (Figure 6.5). To this end, the online

resources include the geometry file to be imported in HEC-RAS (Hec-Ras\Exercise-files\Po_River_SRTM.g01). These cross sections were derived by extracting bed and floodplain bottom levels from the global SRTM topography (Figure 6.5).

SAR data for model evaluation

Between the end of May and the beginning of June 2008 the River Po experienced a low-magnitude flood event. On 1 June at 9:26 a.m., around 1 hour before the peak flow at Cremona, coarse resolution (100 m) SAR imagery was acquired and processed in near real time. The flood image is an acquisition by the ENVISAT-ASAR sensor in WSM and was provided through ESA's Fast Registration system at no cost 24 hours after the acquisition. The ASAR WSM image was processed by Di Baldassarre et al. (2009a) to provide flood extent maps, which can be used to calibrate the model. The online resources of this book also include the SAR-derived inundation width (Hec-Ras\Exercise-files\2008_flood_width.txt).

Simulation of the 2008 flood

Flood events affecting this test site are characterized by broad and slowly varying hydrographs (Di Baldassarre and Claps, 2011). Thus, the 2008 flood can be simulated in steady flow conditions. This allows a strong reduction of the computational cost. Concerning the boundary conditions, at the time of the SAR imagery the hydrometric conditions were (ARPA, 2008):

River discharge at Cremona (inflow) = 5280 m^3 s^{-1}
Water elevation at Borgoforte (downstream boundary condition) = 19.25 m a.s.l.

Specific tasks

6.1 Build flood inundation modelling using the HEC-RAS software and the provided geometry file (cremona_borgoforte.g01). Note that the names of the cross sections indicate the distance (in m) from the downstream end (Borgoforte).

6.2 Run a first steady flow simulation using the boundary conditions given above and homogeneous Manning's coefficient constantly equal to 0.05 m$^{-1/3}$ s in both channel and floodplain. Compare simulated ('Top width (m)' in the 'Profile Output Table') and ASAR-derived inundation width (2008_flood_asar_width.txt) in terms of mean absolute error (MAE; see equation 6.2).

6.3 Run additional simulations by changing the Manning's coefficient between 0.02 m$^{-1/3}$ s and 0.10 m$^{-1/3}$ s and make a diagram of the MAE versus roughness coefficient (Diagram A).

6.4 Analyse the inflow uncertainty. According to Di Baldassarre and Montanari (2009), the uncertainty in the 2008 inflow (river discharge at Cremona at the time of the SAR image) is around ±25%. A simple approach to take this uncertainty into account is to run Monte Carlo simulations using 100 random values of the river discharge at Cremona in the interval [3960; 6600] and the optimal Manning's coefficient (Diagram A). The generation of these random values can be performed with many different tools (e.g. MS Excel, Matlab). Make a diagram of the *MAE* versus the inflow (Diagram B).

6.5 Analyse Diagrams A and B and comment on the results. Discuss the potential role of the other sources of uncertainty not considered here, such as model structure (1D steady flow), inaccuracies of the calibration data and the SRTM topography (Chapter 4).

7   Model outputs

This chapter describes the use of model results in GIS environments and reports the necessary steps to build flood hazard maps. A comparison of deterministic and probabilistic approaches to map floodplain areas is also included.

7.1   MAPPING MODEL RESULTS

Maps are valuable tools to represent the spatial distribution of flood hazard, vulnerability or risk as they provide a more direct and stronger impression than any other form of presentation (Merz et al., 2007). This is why maps are widely used to communicate risk (e.g. Pender and Faulkner, 2010; Hall and Beven, 2011). Moreover, the European Parliament recently published a new European Directive on the assessment and management of flood risks requiring the construction of flood risk maps for all the river basins with significant potential risk of flooding (European Parliament, 2007; Apel et al., 2009).

Floodplain maps can be classified into three categories (e.g. Merz et al., 2007): flood hazard maps, showing the intensity of floods and their associated exceedance probability; flood vulnerability maps, illustrating the consequences of floods on economy, society and the natural environment; and flood risk maps, showing the spatial distribution of the risk, which, for natural disasters, can be defined as the probability that a given event will occur multiplied by its consequences.

Floodplain maps of the inundation area or maximum water levels corresponding to events with a given return period are one of the most common categories of map used to illustrate flood hazard (Bates et al., 2004).

The methodology used to derive floodplain maps with a hydraulic model, such as HEC-RAS, usually comprises the following steps (Merwade et al., 2008):

(1)   The design flood (e.g. the 1-in-100 year flood) is estimated by means of flood frequency analysis or via regionalization or hydrologic modelling in ungauged basins.

(2)   Cross sections for the river and floodplain system are developed. This is done either by ground survey or by extracting elevations along river transects from a DTM (Chapter 4).

(3)   A hydraulic model is first calibrated (and possibly validated) and then executed for the design flood to estimate the water surface elevations at the cross sections developed in step 2.

(4)   The water surface elevations from the hydraulic model are geo-referenced on the DTM, and a water surface (usually a triangular irregular network, TIN format) is created.

(5)   The DTM is subtracted from the water surface to obtain a water-depth map.

(6)   The area with positive water-depth values gives the simulated flood inundation extent.

Regarding the last point, it should be noted that, given the uncertainty in both topography and inundation shoreline (Chapter 4), a small positive threshold depth rather than $h = 0$ is generally used as the test for a pixel's wet/dry state (Aronica et al., 2002). The reason is that, below a certain water depth, model predictions tend to be dominated by topographic noise (Aronica et al., 1998). Also, very shallow water depths may be indistinguishable from surface ponding or exfiltration and not hydraulically connected to the flood inundation. Hence, pixels with simulated water depths below a certain threshold are often treated as dry. The value of such a threshold should depend on the micro-topography and vegetation of the region under study, but typically is between 10 and 20 cm. Aronica et al. (2002) showed that the uncertainty introduced by the selection of this threshold tends to be rather small compared to other sources of uncertainty.

Figure 7.1 shows an example of a floodplain map derived using HEC-RAS. In particular, Figure 7.1 shows flood water levels on the River Po corresponding to a synthetic flood event with a return period of 5 years (1-in-5 year flood; Maione et al., 2003).

The flood hazard map reported in Figure 7.1 was obtained by simulating the 1-in-5 year flood using the HEC-RAS model of the River Po, between Cremona and Borgoforte (see Chapter 6; Figure 6.5). It is important to note that the visualization of HEC-RAS

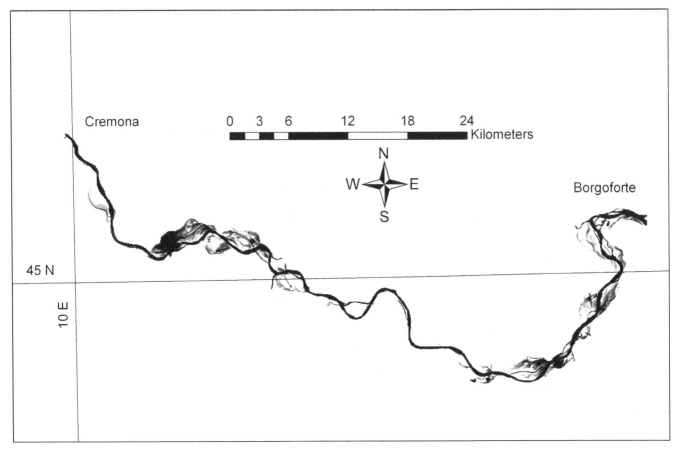

Figure 7.1 Flood hazard map: spatial distribution of maximum water depths (greyscale) corresponding to the 1-in-5 year flood.

results in a GIS environment can be facilitated by the aforementioned tool HEC-GeoRAS (Chapter 5).

Flood hazard maps can be produced using either deterministic or probabilistic approaches (Bates *et al.*, 2004; Merz *et al.*, 2007). Deterministic approaches normally consist of construction of a hydraulic model (Chapter 5), calibration of the model using historical flood data (Chapter 6), use of the best fit model to simulate synthetic design flood events (e.g. the 1-in-100 year flood; Maione *et al.*, 2003), and elaboration of the model results to generate flood hazard maps in a GIS environment. Despite the recent development of complex methodologies for the simulation of the hydraulic behaviour, deterministic predictions of inundation extent using single optimum parameter sets do not take any account of the uncertainties in the modelling process (Bates *et al.*, 2004). This might lead to an incorrect assessment of hazard when the inundation maps are used for other purposes, such as planning decisions for future developments in the vicinity of the floodplain.

Hence, conceiving inundation hazard as a probability has been encouraged more recently (Aronica *et al.*, 1998, 2002; Hall *et al.*, 2005a, b; Romanowicz and Beven, 1996, 1998, 2003; Bates *et al.*, 2004; Pappenberger *et al.*, 2005, 2006). In a probabilistic

approach, floodplain mapping generally consists of construction of flood inundation models (Chapter 5), sensitivity analysis of the model using historical flood data (Chapter 6), and use of the multiple behavioural (acceptable) models to carry out ensemble simulations using an uncertain synthetic design event as hydrologic input (Bates *et al.*, 2004).

This section describes and compares deterministic and probabilistic approaches for floodplain mapping, discussing advantages and disadvantages of the two approaches. The section focuses on the so-called 1-in-100 year flood inundation map, widely used in floodplain mapping and supporting land use and urban planning. The 1-in-100 year flood inundation map shows the floodplain area that is supposed to be flooded, on average, at least once every 100 years. This map is usually obtained by mapping the results of a hydraulic model where the 1-in-100 year flood (defined as the river discharge with a return period of 100 years) is used as hydrologic input (Bates *et al.*, 2004). A hundred years is one of the typical recurrence intervals used for design purposes (Maione *et al.*, 2003; Merz *et al.*, 2007). For example, the 1-in-100 year flood is used by the Environment Agency of England and Wales to derive indicative floodplain maps showing areas at risk from flooding (e.g. Bates *et al.*, 2004).

Figure 7.2 Contour of the 2006 flood extent derived from ERS-2 SAR image (black line) and deterministic floodplain mapping of the 1-in-100 year event (grey area).

7.2 DETERMINISTIC FLOODPLAIN MAPPING

This section illustrates an example of deterministic floodplain mapping by referring to a river reach of the Lower Dee, UK. To this end, the LISFLOOD-FP code (Bates and De Roo, 2000; Chapter 5) was used. The LISFLOOD-FP model was calibrated by referring to an ordinary (return period equal to around 2 years) flood event that occurred in December 2006. Hydrometric data observed at Environment Agency gauging stations are used as boundary conditions (Di Baldassarre *et al.*, 2009b, 2010b). The model sensitivity to floodplain friction was found to be small at this test site. This is likely to be because the floodplain here acts predominantly as a storage area characterized by low-velocity flow. As the total frictional force is proportional to the square of the velocity, floodplain roughness will be relatively unimportant in such situations (Di Baldassarre *et al.*, 2010b). Thus, LISFLOOD-FP was run with a uniform floodplain Manning's coefficient of $0.10 \text{ m}^{-1/3}$ s and the model calibration was performed by running simulations by varying the Manning's channel coefficient in the range $0.02–0.10 \text{ m}^{-1/3}$ s.

The results of each model are compared to the flood extent map derived from an ERS-2 SAR image (Figure 7.2) in terms of measure of fit F_2 (Chapter 6). The outcome of the sensitivity analysis is illustrated in Figure 7.3, showing the model response to changes in Manning's coefficients. By analysing Figure 7.3,

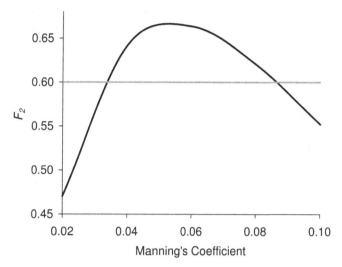

Figure 7.3 LISFLOOD-FP response to changes in Manning's coefficient (black line) and performance of a 0D model, i.e. simple planar model (grey line). In the example of a GLUE application, the models performing worse than the 0D model were rejected (Di Baldassarre *et al.*, 2010b).

one can observe that the 'best fit' model is characterized by an F_2 value equal to 0.64.

To generate the deterministic 1-in-100 year floodplain map, the design event (i.e. the 1-in-100 year flood) was used as hydrologic input of the 'best fit' model. The 1-in-100 year flood was estimated by the Environment Agency (2003), applying the FEH statistical method (Institute of Hydrology, 1999). Then, the results of the best fit model were elaborated in a GIS environment to derive the deterministic 1-in-100 year flood inundation map (Figure 7.2). The result is a precise inundation map, but one that is potentially inaccurate if the assumption regarding stationarity of parameter values between events does not hold.

7.3 PROBABILISTIC FLOODPLAIN MAPPING

This section describes an example of probabilistic floodplain mapping, using the same case study. To this end, the GLUE (Beven and Binley, 1992; Aronica *et al.*, 2002; Horritt, 2006; Chapter 6) framework was applied. More specifically, the results of the sensitivity analysis (Figure 7.3) allowed the selection of a subset of behavioural models (Beven and Binley, 1992) that were then used to simulate the 1-in-100 year flood event (Bates *et al.*, 2004). This study rejected all the models that underperform the simple planar model (0D model; Chapter 5), which consists of a linear interpolation of the measured water levels to derive water surface and then inundated area (Horritt *et al.*, 2007; Apel *et al.*, 2009). Thus, the performance of a simple planar model (equal to about 0.60; Figure 7.3) was used as threshold to distinguish between behavioural and non-behavioural LISFLOOD-FP models (Figure 7.3). It is important to note that, although the simple planar model may sometimes reproduce historical flood events with acceptable performance (e.g. Horritt *et al.*, 2007; Di Baldassarre *et al.*, 2010b), this cannot be easily utilized to predict the 1-in-100 year flood extent where measured water levels are not available. In addition, the planar method has a number of disadvantages such as (Apel *et al.*, 2009): there is no volume control of the floodplain inundation; the effects of hydraulic structures cannot be modelled; and the output of hydrologic models (river discharge) cannot be used as input. Hence, for the simulation of design events, such as the 1-in-100 year flood, the use of hydraulic models is usually recommended.

In the probabilistic approach, all the behavioural models were used to simulate the 1-in-100 year flood. The results were then combined to produce a probabilistic 1-in-100 year flood inundation map, conditioned on the 2006 data, in a GLUE framework. In particular, each simulation i was attributed a likelihood weight L_i in the range [0,1] according to the values of F (Figure 7.3):

$$L_i = \frac{F_{2,i} - \min(F_{2,i})}{\max(F_2) - \min(F_2)} \qquad (7.1)$$

where $\max(F_2)$ and $\min(F_2)$ are the maximum and minimum measures of fit found throughout the ensemble. Then, given the simulation results for the jth computational cell of w_{ij} equal to 1 for wet and w_{ij} equal to 0 for dry, the uncertain 1-in-100 year flood inundation map can be produced by evaluating:

$$C_j = \frac{\sum\limits_i L_i w_{ij}}{\sum\limits_i L_i} \qquad (7.2)$$

where C_j indicates a weighted average flood state for the jth cell.

Figure 7.4 shows the uncertain 1-in-100 year flood inundation map obtained by combining the results of these numerical simulations. While C_j is not a probability in the strict sense (Montanari, 2007), it does range between 0 and 1, and reflects the likelihood of inundation at that point for a 1-in-100 year flood event (Horritt, 2006; Di Baldassarre *et al.*, 2010b).

The map reported in Figure 7.4 does not take into account the uncertainty in the 1-in-100 year flood event magnitude. In fact, the estimation of the design flood is affected by errors caused by the choice of a probabilistic model as well as the parameterization of the model itself (e.g. Laio, 2004; Mitosek *et al.*, 2006; Viglione *et al.*, 2007). In order to produce an additional uncertain 1-in-100 year flood inundation map, which also takes into account the uncertainty in the design flood, a simplified approach was applied referring to previous findings of Di Baldassarre *et al.* (2009d). According to these findings a probabilistic flood inundation map (Figure 7.5) was derived by applying the GLUE procedure (see above) and using as boundary condition for the ensemble simulation a random value normally distributed with average equal to the estimated 1-in-100 year discharge and standard deviation equal

Figure 7.4 Probabilistic floodplain mapping of the 1-in-100 year event (5th, 50th and 95th weighted percentiles; greyscale). This flood inundation map explicitly considers uncertainty in model parameters.

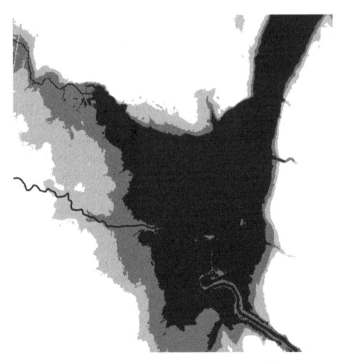

Figure 7.5 Probabilistic floodplain mapping of the 1-in-100 year event (5th, 50th and 95th weighted percentiles; greyscale). This flood inundation map explicitly considers uncertainty in model parameters and design flood estimation.

to ±15% (Di Baldassarre *et al.*, 2009d). This 15% is expected to be an optimistic estimate of the uncertainty in design flood estimation as this simplified estimate does not take into account the additional uncertainty induced by the non-stationarity that might arise from land-use and climate changes (Chapter 1) nor does it consider the inaccuracy of river discharge data (i.e. time series of annual maximum discharge; Chapter 4) used for the flood frequency analysis. In fact, a number of studies showed that this type of uncertainty may be very high for data referred to high-flow conditions when stage–discharge rating curves are extrapolated beyond the measurement range (e.g. Clarke, 1999; Petersen-Øverleir, 2004; Pappenberger *et al.*, 2006). For instance, Pappenberger *et al.* (2006) pointed out that uncertainty affecting recorded river discharge data might be equal to about 20%, even before the observed values are used to extrapolate to low-frequency events.

By analysing and comparing Figures 7.4 and 7.5, one can observe that, despite the simplified and optimistic approach applied to take into account the uncertainty in the 1-in-100 year peak discharge, this additional source of uncertainty strongly increases the uncertainty in floodplain mapping. Furthermore, it is important to note that the uncertainty in design flood estimation is expected to be much higher for synthetic events of higher magnitude (e.g. 1-in-500 or 1-in-1,000 year events).

7.4 DETERMINISTIC VERSUS PROBABILISTIC

Floodplain mapping is affected by many sources of uncertainty including: observational errors in hydrologic inputs, inaccurate definitions of the topography of rivers and floodplains, inappropriate model structures (or inability of capturing some processes, such as sediment transport, evapotranspiration, groundwater integrations, etc.), the approximate parameterization, inaccuracy of data used for model evaluation, in addition to the non-stationarity that might arise from changes in climate and/or land use.

However, floodplain mapping is often based on a single deterministic prediction of the flood inundation area (Merwade *et al.*, 2008; Di Baldassarre *et al.*, 2010b). In this chapter, a case study is used to critically show and discuss deterministic and probabilistic approaches to derive flood inundation maps. Theoretically speaking, visualizing flood hazard as a probability is a more correct representation of the subject since deterministic predictions of inundation extent and design discharge, which use the single best fit model and best estimate peak discharge, might misrepresent the uncertainty in the modelling process and give a result that is spuriously precise (Beven and Freer, 2001; Bates *et al.*, 2004; Beven, 2006). For instance, by comparing the SAR-derived 2006 inundation extent and the deterministic 1-in-100 year flood inundation map (Figure 7.2) one can observe that, although the latter

is overall wider than the former, there are some floodplain areas that are outside the deterministic 1-in-100 year flood inundation map but were observed as inundated during the 2006 flood event. Given that the 2006 flood event had a return period much lower than 100 years (see above), this demonstrates that deterministic flood inundation maps are only spuriously precise.

Moreover, deterministic approaches are based on the hypothesis that hydraulic models, once calibrated (and, in some rare cases, also validated) using historical flood data, are able to correctly predict flood events of different magnitude. For instance, referring to the application example illustrated in this chapter, it is assumed that the LISFLOOD-FP model, once calibrated using the 2006 flood extent, gives a reliable simulation of the inundation processes driven by the 1-in-100 year flood event. This assumption is very debatable. In fact, a number of studies (e.g. Aronica *et al.*, 1998; Horritt and Bates, 2002; Romanowicz and Beven, 2003; Horritt *et al.*, 2007; Di Baldassarre *et al.*, 2009a) have shown that a well-calibrated flood inundation model may perform very poorly when it is used to predict different events. In particular, these studies showed that the effective roughness coefficients may be different when evaluated for flood events of different magnitude, even when uncertainty in calibrating roughness coefficients is allowed for. Therefore, a flood inundation model, calibrated on a historical event, may give a poor prediction of a synthetic design event, especially if this is characterized by a different magnitude (Di Baldassarre *et al.*, 2009a). Unfortunately, these effects of non-stationarity of friction parameters can only be partially decreased using physically based fully 2D hydraulic models (Di Baldassarre *et al.*, 2010b).

In contrast, probabilistic approaches are less sensitive to the non-stationarity of model parameters, as the use of multiple behavioural models in prediction, rather than a single best fit model, helps to reduce these effects (Bates *et al.*, 2004). Furthermore, using different types of data in the calibration process might lead to different outcomes in prediction (both deterministic and probabilistic) due to error and bias in the calibration data. For instance, SAR data typically allow a trade-off between channel and floodplain friction and can identify as acceptable some parameter sets that are physically implausible, while if one had some spot high water marks a different prediction would be achieved (see also Chapter 6).

In terms of modelling, advanced deterministic approaches to derive flood hazard maps theoretically require the implementation of sophisticated numerical models characterized by complex model set-up (e.g. construction of unstructured computational meshes). In floodplain mapping, the expertise and the time required to build and set up the model are very significant (while for real-time flood inundation forecasting the computational time is the main issue). For instance, in order to generate flood hazard maps for all the river basins with significant potential risk of flooding in Europe (as recommended by the Directive of the European

Parliament, 2007) it is obviously easier to implement simplified models rather than more sophisticated ones. In fact, this is what is commonly done by the stakeholders: use of simplified models, but (unfortunately) in a deterministic approach.

In contrast, probabilistic approaches can be based on simple flood inundation models. One can argue that probabilistic approaches require extensive sensitivity analysis and ensemble simulation. However, uncertainty estimation is not necessarily difficult to implement. For example, the methodology described in this chapter (GLUE), which is probably the most used uncertainty estimation method in hydrology research (Montanari, 2007), is relatively simple to apply. Nevertheless, environmental agencies, river basin authorities and engineering consultancies hardly ever apply probabilistic approaches for floodplain mapping. This is partly due to the fact that the transfer of relative know-how from scientists to end-users is still difficult (Montanari, 2007). In addition, it important to note that part of the scientific community is still reluctant to embrace probabilistic approaches (Pappenberger and Beven, 2006). Figures 7.2 and 7.4 may help to explain this reluctance. It is commonly believed that people would find the deterministic map (Figure 7.2) more straightforward than uncertain flood inundation maps (Figures 7.5 and 7.6), as it seems to better reflect the common landscape view of river, floodplain and the rest of the territory. However, most people recognize the fuzziness that exists between these landscape elements as well as the fact that not all the floodplain areas are exposed to the same flood hazard. In addition, there are scientists who believe that decision-makers would prefer deterministic binary maps over probabilistic maps (roughly speaking, for setting floodplain planning rules, one might wonder: what shall I decide for an area classified as 0.3181?). Nevertheless, it seems that this is mainly due to the fact that, so far, uncertainty estimation in hydrology suffers from the lack of a coherent terminology and a systematic approach (Montanari, 2007) and mature guidance on methods and applications does not exist (Pappenberger and Beven, 2006; Di Baldassarre et al., 2010b).

7.5 CONCLUSIONS AND PERSPECTIVES

Floodplain mapping is often based on the deterministic application of hydraulic models to produce spuriously precise flood hazard maps without proper consideration of the intrinsic uncertainty. Such an approach should be revised for different reasons. In fact, to produce a scientifically justifiable deterministic map the most physically realistic model should be utilized. Unfortunately, these sophisticated models are usually characterized by complex model set-up (e.g. mesh construction) and require a high level of expertise and time to construct. Moreover, this chapter showed that, even when derived using physically based models, deterministic maps are only spuriously precise given: (i) the frequently observed tendency for non-stationarity of parameter values with changing event magnitude and timing and (ii) fundamental uncertainties arising from extreme event statistical analysis used to determine the magnitude of design events such as the 1-in-100 year flood. In fact, model structures are far from being perfect and input/calibration data are often affected by non-negligible errors. In addition, the design flood used to produce flood hazard maps corresponding to a certain return period (e.g. 1-in-100 year event) is affected by significant uncertainty due to the use of imprecise probabilistic models to infer inaccurate river flow data and non-stationarity that might arise from changes in land use and climate. Hence, visualizing flood hazard as a probability seems to be a more correct representation of the subject. In order to assist the diffusion of probabilistic approaches for floodplain mapping, clear methods and applications need to be established.

7.6 EXERCISES

7.1 Illustrate the common steps required to build floodplain maps using a 1D hydraulic model.
7.2 Explain why deterministic approaches for floodplain mapping often lead to products that are only spuriously precise.
7.3 List the main sources of uncertainty usually affecting floodplain mapping.
7.4 This chapter described a probabilistic approach to produce flood inundation maps based on the GLUE framework. Make a literature search to find alternative methods and highlight advantages and disadvantages of these methods compared with GLUE.
7.5 Explain why probabilistic floodplain maps might prove difficult to communicate and therefore floodplain mapping is still made without explicit consideration of the intrinsic uncertainty.

Part III
Applications

8 Urban flood modelling

Contributing authors: Jeffrey C. Neal, Paul D. Bates and Timothy J. Fewtrell

8.1 INTRODUCTION

The global population is becoming increasingly urbanized, and this, along with rising total global population, sea level rise and possible future climate change, is leading to an unprecedented increase in urban flood risk. Urban systems are also potentially much less resilient than rural ones because of the complex interdependency of urban infrastructure. There is thus a clear need to be able to model flood risk in urban areas and this presents a set of distinct challenges for hydraulic modellers, which have only recently begun to be addressed. In this chapter we review the latest developments in the science of urban flood inundation modelling. First, we consider the data necessary to build and evaluate urban hydraulic models (Section 8.2) and show how this involves an order of magnitude increase in data and model resolution compared to that required for rural test cases. Following this general analysis we give an example of flood model development for the city of Carlisle in the UK (Section 8.3), where an exceptional data set has been assembled that enables rigorous development and testing of urban hydraulic models. We summarize the lessons learned from these recent studies in Section 8.4.

8.2 REQUIREMENTS FOR HYDRAULIC MODELLING OF URBAN FLOODS

A number of authors have highlighted the significant data requirements for successful flood modelling studies in rural areas (Bates, 2004b; Hunter et al., 2005a). The challenges faced by a shift to the consideration of urban environments are considerable and principally relate to data, but there is currently also a need for further study of the fluid dynamics of urban flood flows at the field scale (Hunter et al., 2008). The data requirements can be conceptualized as two distinct but inherently linked units, the data needed to build numerical models and the data needed to evaluate these numerical models.

8.2.1 Building hydraulic models

This section discusses the three essential data sources required for numerical modelling of urban inundation, namely topography, surface friction and boundary conditions, with specific reference to the issues faced in urban areas.

As flood flows are typically shallow and follow complex flow paths, topography is an essential component of any numerical inundation model. Parameterization of topographic data has been explored extensively in rural areas (e.g. Bates et al., 1998, 2003; Cobby et al., 2003) but more recent studies in urban areas have begun to emerge (e.g. Mason et al., 2007; Néelz and Pender, 2006; Neal et al., 2009a; Mignot et al., 2006). Mason et al. (2007) note that high spatial frequencies of elevation change are characteristic of urban topography. From a hydraulic viewpoint, these have a significant effect on flood wave propagation and storage (Mignot et al., 2006; Yu and Lane, 2006) and from a modelling standpoint, the varying shapes and length scales determine the grid resolution of any model (Mark et al., 2004). In fact, the surface drainage network may be approximated as a series of 1D channels (i.e. roads) connected at storage areas (i.e. road junctions, squares) and thus modelled as such (Braschi et al., 1989). However, this assumes that the flow paths are known a priori and that open areas act purely as storage rather than as a mode of conveyance. Hence 2D modelling approaches, which tend to better represent momentum transfer through urban areas, are becoming increasingly popular. Since these models require two-dimensionally distributed topographic data for parameterization, the development of urban flood models (along with flood modelling in general) has been enhanced greatly by technological advances in digital elevation modelling (DEM) derived mostly from laser altimetry (LiDAR), which can produce data at sub-metre-scale accuracy and precision. Sonar surveys of channel bathymetry (e.g. Eilertsen and Hansen, 2008; Horritt et al., 2006) and the use of digital mapping data (e.g. Mason et al., 2007) have further reduced uncertainties associated with topographic data sets.

Traditionally, 2D modelling techniques have been limited not only by the sparsity of topographic data, but also by computer processing power. Recent advances in computing technology (e.g. graphics processing units (GPUs) and high-performance computing (HPC)) have relaxed these constraints for some applications such that the high-resolution data sources can be exploited to something approaching their full potential. Nevertheless, not all models and applications can be scaled to the <1 m resolutions attainable from LiDAR, meaning intuitive and physically based methods for aggregating such data to scales at which the current suite of numerical models are computationally feasible and efficient for engineering and planning applications are required. However, aggregation to coarse model grid scales generally assumes that the governing equations still hold and that effective parameters can be found appropriate to the model scale (Beven, 1995). With reference to flood modelling, if the assumption that, on a large scale, a flood wave is still a slowly propagating, gradually varying wave, then the governing equations may remain unchanged. In practice, Lane (2005) notes that for topographic parameterization, a change of model scale necessitates a change in the degree to which topography is parameterized implicitly (i.e. as frictional resistance) rather than represented explicitly. However, the issues of scale and aggregation of topographic data in urban applications are largely unexplored in urban hydraulic modelling (the study of Fewtrell et al., 2008 being an exception here).

Urban development typically involves the removal of trees, the replacement of soils and vegetation with impervious surfaces, and the replacement of the natural drainage system with a network of storm sewers (Nelson et al., 2009). These impervious surfaces act on the hydrology of the urban environment to reduce interception of rainfall by the canopy and infiltration into the subsurface and thus increase the fraction of rainfall that becomes runoff. However, there are also impacts on surface water hydraulics as overland flow velocities are substantially faster on smooth surfaces (e.g. concrete, asphalt).

In rural areas detailed LiDAR return information can be used to inform friction parameterization in 2D numerical flood models (e.g. Mason et al., 2003), which Bates (2004a) notes may lead to the prospect of spatially distributed grid-scale effective parameters and thus a reduced need for calibration of hydraulic models. Applying a similar technique in urban environments may be possible with detailed land-use information from digital mapping data sets (e.g. MasterMap®). However, such a method assumes that the aerial average friction is the only significant frictional resistance to flow at the grid scale, which may not be the case (Lane, 2005). However, Beven (2006) notes that friction values at coarse grid scales may not be physically based, but rather may be truly effective parameters that cannot be easily determined a priori. Furthermore, the use of land-use classifications and empirically determined values from literature (e.g. Chow, 1959) to assign friction values may be meaningless as most friction formulations (e.g. Manning's n, Chezy's C) were derived for natural rivers and

should not be applied outside this context (Lane, 2005). In addition, this is only an appropriate method if the basis of derivation of the floodplain friction values uses the same assumptions as the model being applied to the floodplain. Therefore, although topographic and topological data sets may provide guidance for the derivation of friction values, these values are inherently calibration parameters, and where possible should be treated as such in any modelling framework.

Boundary conditions for hydraulic modelling of floods, whether 1D or 2D, are generally specified as flow or water stage hydrographs derived from gauging stations at the top (and sometimes bottom) of the modelled reach. However, gauging stations are often designed with water resource management or flood warning, rather than hydraulic modelling, in mind. As such, during flood events these gauges often operate outside the designed measurement range, introducing significant uncertainties to these data. Furthermore, as typical gauge spacing in the UK is 10–60 km or more apart, few such data are available (Bates, 2004a). Moreover, urban areas subject to surface water flooding due to excess rainfall and which are disconnected from the main river network are entirely ungauged. Uncertainties in input data, when subject to extrapolation to larger events or into the future, may generate significant deviations in model results that can negate any predictive ability (Oreskes et al., 1994). Furthermore, the assumption that present observations are indicative of future conditions is not guaranteed as natural systems are dynamic (Oreskes et al., 1994). The alteration of gauging station reaches and flow dynamics by vegetation, floodplain development and sediment transport represents practical limitations to current gauging station data sets. Nevertheless, gauge data are typically the most accurate data sets on river flows available and are widely used in both rural and urban applications.

8.2.2   Assessing urban flood models

The combination of uncertainties in parameter values and initial and boundary conditions initiates an uncertainty cascade (Pappenberger et al., 2006) that propagates to model predictions of water depths and consequently to estimates of flood damage. Until recently, validation data for hydraulic models have largely been bulk measurements (stage or discharge at points on the river network) representing the spatially aggregated catchment response. However, flood inundation modelling is a spatially and temporally distributed problem that requires distributed, rather than lumped, observational data to constrain and validate model predictions (Bates, 2004a) because many parameter sets tend to simulate lumped observations with similar accuracy (e.g. the model parameters are equifinal, given the observations). To put it another way, bulk flow measurements represent an aggregate catchment response to that point, and thus evaluating hydraulic models with these data can lead to a wide range of plausible models and parameter sets. For any given model, many different combinations of flow conditions and grid-scale effective parameter values may lead to the same aggregate catchment response but give different spatial predictions and, thus, process inferences. In fact, replication of aggregate catchment response often only requires single values of model parameters spatially lumped at the catchment scale (Bates, 2004a). As such, stage and discharge data are unlikely to provide a sufficiently rigorous test for competing model structures (Hunter *et al.*, 2005a) and, indeed, render model parameterizations indistinguishable from each other (Beven, 2002). Nonetheless, flow records have proved their utility in testing the wave routing behaviour of flood models and have been shown to be replicable by even the simplest of numerical schemes (Horritt and Bates, 2002).

The integration of remotely sensed imagery with flood models (e.g. Horritt *et al.*, 2007; Schumann *et al.*, 2007b) and the use of spatially distributed point measurements (e.g. Hunter *et al.*, 2005a; McMillan and Brasington, 2007; Neal *et al.*, 2009a) provides large distributed data sets with which to evaluate competing model structures and parameterizations. Using such data is not without problems, a consideration of the observations often highlights the mismatch between the nature of variables used to run and evaluate a model and the nature of the observed variable (Freer and Beven, 2005). At the local point scale (e.g. a surveyed water level measurement compared to the free surface elevation predicted at the effective model grid scale), this difference arises as a result of scale, heterogeneity, non-linearity and incommensurability effects, so that the predicted variable is not the same quantity as that measured (Beven, 2006), and may not even be indicative of the natural phenomenon (Oreskes *et al.*, 1994). Oreskes *et al.* (1994) further note that observations and measurements of both independent and dependent variables are laden with inferences and assumptions attributed to the environmental modeller. In practical terms, what is perceived as a maximum water level mark may purely be the level at which water remained ponded during floodwave recession. Similarly, ponded water may deposit wrack marks that may be incorrectly interpreted as maximum flood extents. Given the noise in observations (spatially and/or temporally) used to evaluate model predictions (Beven, 2006), model states will inevitably be both equifinal and indistinguishable. Furthermore, Hunter *et al.* (2005a) note that there is a trend in environmental modelling to ignore the errors and uncertainties associated with field measurements due to the difficulties in collecting these data. However, errors and uncertainties in these data may have a significant impact on the predictive ability of flood models or values of effective parameters estimated within distributed models, depending on the modelling application.

Synoptic-scale maps of floods processed from remotely sensed data provide wide-area, spatially distributed and spatially and temporally discrete information on flood extents. Such data have been extensively used and evaluated for constraining hydraulic models on rural reaches (see Horritt and Bates, 2002; Hunter *et al.*, 2005a; Schumann *et al.*, 2007b) where topographic variation has a fractal nature at large spatial scales. However, significant elevation changes on short spatial scales in urban areas and the channelized nature of many urban floods requires that remotely sensed imagery of flooding capture the detailed variation in flood extent between urban structures. In fact, the resolution requirements for remotely sensed imagery used to evaluate urban flood patterns (~1–2 m) far exceed current satellite capabilities (~20 m ground resolution) and the availability of airborne data is limited. Furthermore, even with future advances in satellite technology (e.g. TerraSAR-X at ~3 m ground resolution), problems of detecting building/ground/water transitions will still remain, as complex radar returns from these surfaces will make flood delineation problematic in urban areas. As a consequence of errors in observational data and the mismatch of scales in remotely sensed imagery, Beven (2006) suggests that modellers can (or should) only look for application-specific consistency between modelled and observed data.

8.3 TEST CASE

The discussion above has highlighted the wide range of high-resolution data required to build and assess hydraulic models of urban floods. However, the sparse availability of all these data significantly restricts the sites at which urban flood risk can be analysed in detail. The city of Carlisle in the UK (Figure 8.1) is one of very few sites where sufficient data sets are available to both build the model and evaluate its performance during a significant flood event.

8.3.1 Site and event description

In January 2005, the city of Carlisle in Cumbria, UK, experienced substantial flooding as a result of water levels approximately one metre above the 1822 level, the previous highest recorded flood level in Carlisle. The city is situated at the confluence of one major river (River Eden) and two significant tributaries (Rivers Petteril and Caldew) with a combined catchment area of ~2,400 km^2. The Petteril and Caldew rivers are both subject to rapid flood response as a result of the steep upper regions of the catchments (Clarke, 2005). The majority of the catchment is rural with the major urbanization concentrated around Carlisle, which is where this test case is situated. High flows are generally contained by the defence structures, although these defences are estimated at only providing protection up to the 1-in-70 year event.

Initial estimates suggested the 2005 flood event was in the region of a 1-in-250 year flood event on all three rivers, but

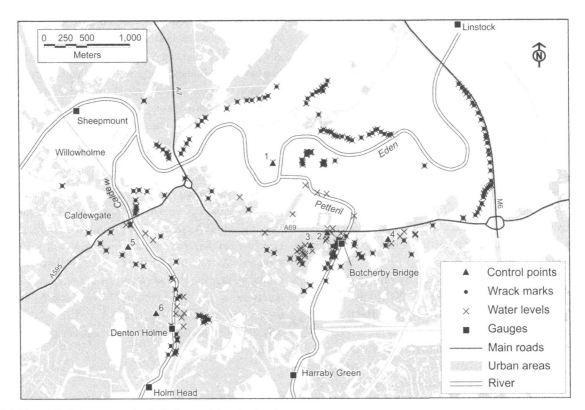

Figure 8.1 Map of Carlisle test case site, including model evaluation data.

subsequent investigations have found the event to be a 1-in-150 year event on the River Eden and a 1-in-100 year event on the Caldew and Petteril rivers (Clarke, 2005). The flooding was largely caused by high river levels as a result of almost continuous heavy rainfall from January 6th to 8th. The storm event began on January 6th and was accompanied by gale force winds on January 7th and 8th. The River Eden catchment received up to 175 mm of rain in the 36-hour period (Day, 2005). Furthermore, the wet antecedent catchment soil conditions and the associated full lakes offered little storage capacity causing rapid runoff into the rivers. The resulting river flows were up to 1,600 m³ s⁻¹ on the River Eden in Carlisle city centre. These high river flows overwhelmed a number of defences in the Carlisle region, causing widespread flooding throughout the city.

The flooding affected approximately 6,000 residents and 3,500 homes (of which approximately 1,900 properties were directly flooded) and 60,000 homes were cut off from electricity supplies (Day, 2005). Furthermore, the fire station, police station, bus depot and football ground were severely affected by the flooding, with the bus depot forced to scrap the entire fleet. Clarke (2005) estimates the monetary damage from the flood to be ~£500 million.

In October 2004, the Environment Agency published a revised Flood Risk Management Strategy for Carlisle and the Lower Eden for public consultation, in order to cope with the significant flood risk in the area. The scale of the January 2005 floods prompted a rapid reappraisal of the proposals outlined in 2004 to ensure

lessons are learnt from the largest event in recent history (Clarke, 2005).

8.3.2 Data availability and collection

The January 2005 Carlisle flood event provides a unique opportunity to evaluate the data sources available for setting up distributed flood models and assessing model accuracy for urban applications. Data for model set-up is in the form of LiDAR elevation data, river cross sections and river discharge time series. Field measurements of high water marks and flood extents form the basis of model evaluation schemes. This is representative of data that would be routinely gathered before, during and after flood events in the UK.

Airborne scanning laser altimetry data (LiDAR) at metre spatial resolutions are becoming increasingly available (Marks and Bates, 2000) for the generation of digital surface and terrain models for urban areas. Mason et al. (2007) detail the development of a LiDAR postprocessing framework incorporating digital map data and pattern recognition techniques to construct a DTM of the ground surface and a DEM incorporating buildings and vegetation of the area surrounding Carlisle. Figure 8.2 shows the DEM constructed using LiDAR data flown by the Environment Agency in 2003, while Figure 8.3 shows a land-use map derived from a MasterMap® digital map. The LiDAR segmenter is an extension of a similar method developed for rural areas in order to overcome a number of problems inherent in elevation data of urban areas

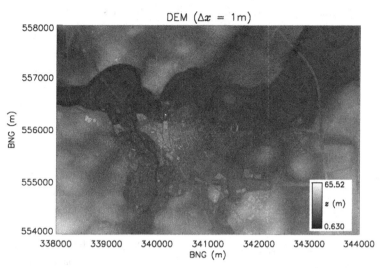

Figure 8.2 Digital elevation model (DEM) of Carlisle site from LiDAR segmented using MasterMap® data (Neal *et al.*, 2009c).

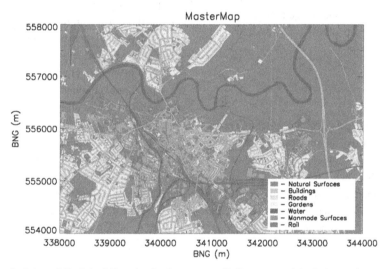

Figure 8.3 MasterMap® topological data of Carlisle delineating land-use types. © Crown copyright Ordnance Survey. All rights reserved.

(Mason *et al.*, 2007). The area covered by the LiDAR survey is approximately 6 × 4 km.

As discussed above, flood inundation models are driven by discharge or water level measurements as upstream, downstream and/or internal boundary conditions. During the Carlisle flood, significant out-of-bank flows at both rated and unrated gauging stations resulted in substantial uncertainty surrounding flow estimates for the event. For hydraulic modelling purposes, the presence of a number of level-only gauges around Carlisle complicates the delineation of a model domain, although the gauges internal to the domain can act as important tools for model calibration and validation. On the River Eden, the lack of a rated gauge upstream of the area of interest and the known problems with the rating curve above 7.0 m water depth at the Sheepmount gauge required significant attention prior to any hydraulic

modelling (Neal *et al.*, 2009a). The rating curve for each gauge was assessed, and in cases where there was doubt over the measurements the sections were re-rated using a full 2D unstructured grid model SFV (Horritt *et al.*, 2010). This resulted in the hydrographs for the 2005 flood event on the Eden, Petteril and Caldew rivers shown in Figure 8.4.

In order to exploit this opportunity to increase our understanding of flooding in the urban environment, a post-event mapping survey of water levels in Carlisle was undertaken. Although undertaking a survey directly after the event does not capture the dynamics of water levels, trash lines and wrack marks left behind by the flood waters are temporary features that record the locations of flood waters. Using a differential global positioning system (DGPS), the [x, y, z] location of individual wrack lines and water marks was collected throughout Carlisle city centre. This data

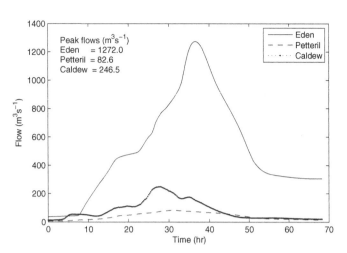

Figure 8.4 Event hydrograph used in this scenario derived from observations of flooding in July, 2002 (Neal *et al.*, 2009c).

set of ∼75 points was combined with the Environment Agency post-event mapping data set of ∼500 points (see Figure 8.1) and represents one of the largest data sets of urban flood extents and water heights.

8.3.3 Analysis of buildings

As stated above, the computational cost of a 2D model increases with resolution, thus an analysis that informs on the resolution needed to capture the dominant flow processes expected in any particular application might be beneficial. In urban and rural areas the topography is usually the principal factor controlling shallow water flows, so an analysis of the length scales over which surface elevations change with similar magnitude to the expected flow depths is needed. In rural areas floodplain gradients are shallow, to the extent that model resolutions of order 10^1–10^2 m are often adequate. However, in urban areas the greater density of man-made structures such as buildings and walls can dramatically decrease the length scales over which changes in topography greater than flow depth occur.

An analysis of digital map data of building footprints was therefore used to help inform the design of the hydraulic model that was used to simulate flood inundation. The digital map data described the locations of building footprints to the nearest metre over the entire city and floodplain. These data were used to calculate metrics on the distribution of building sizes in the city and, perhaps more importantly, the smallest separations between buildings and their neighbours, which might control flow conveyance. Figure 8.5 plots the resulting distribution from this analysis and also includes information on building footprints and longest building axis. The data indicate that the majority of building separations are of the order of 1–10 m and that any gridded model that captures most of these gaps would need to be at or below this resolution.

8.3.4 Results

Floodplain inundation can be simulated using models of one, two or three spatial dimensions. However, 2D models or hybrid 1D and 2D models are typically assumed to be the most suitable in urban areas because the 2D representation allows for the often complex flow paths through the urban environment to be explicitly considered, while assuming averaging velocity in the vertical dimension is adequate. The problem with modelling in two dimensions is the trade-off between model resolution and computation time, particularly when the model uses a structured grid. This occurs not just because the number of locations where computation is needed increases with resolution, but also because the time-steps at which the models can be run also decrease with resolution.

The analysis of building separations in the previous section illustrated how grid spacing down to a few metres is needed to represent 90% or more of the smallest gaps between buildings in Carlisle. It is possible to run some of the more efficient 2D hydraulic models at this resolution over entire cities (e.g. Neal *et al.*, 2011). However, this may not be practical for applications where many simulations are required, meaning a more pragmatic approach is required. Here the resolution of the model has been set at 5 m in an attempt to capture the majority of shortest building separations, although studies have looked at porosity-based approaches to representing subgrid-scale topography (e.g. Yu and Lane, 2006). Researchers have also examined the sensitivity of urban inundation models to resolutions (e.g. Fewtrell *et al.*, 2008).

The hydraulic model used in this example is the LISFLOOD-FP hydraulic model of Bates *et al.* (2010), which has been applied to a number of test cases including Neal *et al.* (2011). This model can be set up to represent the river system as either a 1D channel coupled to a 2D floodplain or as an entirely 2D domain, where the channel bed elevations are represented by the DEM. Here the latter 2D approach was used. The Manning's coefficient needed by the model can be estimated, with some unknown uncertainty, from the land cover of the model domain. However, the model used here was calibrated using the maximum water surface elevation observations from the 2005 event because this resulted in more accurate simulations than using the first-guess Manning's coefficients for this reach. The simple deterministic calibration approach from Neal *et al.* (2009a) was used here to estimate a lumped Manning's coefficient for the channel, where the model performance was measured by the root mean squared error (RMSE) between observational data and model simulations. This is a simple yet efficient approach, although numerous more elaborate calibration methodologies that account for model and observational uncertainty have been developed (Pappenberger *et al.*, 2005, 2006), which could provide a more comprehensive analysis of model performance.

The resulting deterministic simulation of maximum inundation depth, with a RMSE of 0.24 m, is shown in Figure 8.6, along with the data used for model evaluation. This RMSE value is likely

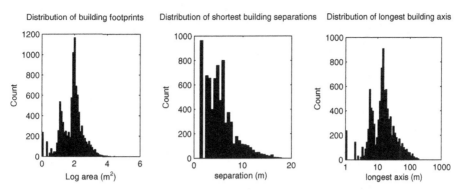

Figure 8.5 Plot of the distribution of smallest gaps between buildings for the city of Carlisle, UK (Neal *et al.*, 2009c).

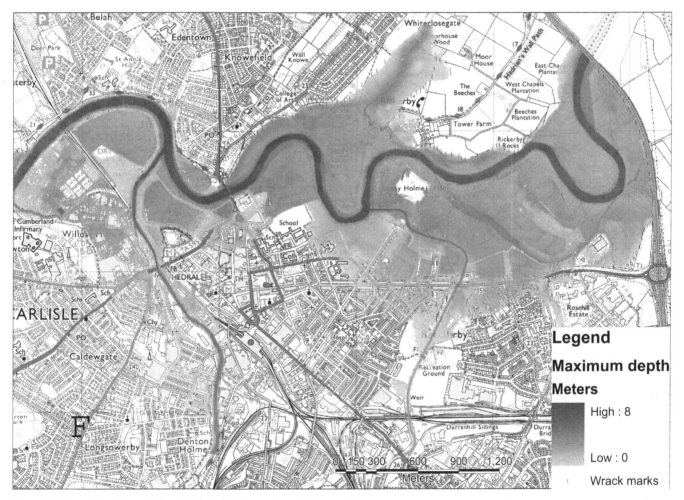

Figure 8.6 Map of maximum simulated depth and observations of flood extent.

within the error in the observed data, which were estimated to be between 0.3 m and 0.5 m by Horritt *et al.* (2010) and Neal *et al.* (2009a). The model does not recreate the flood extent on the upper sections of the River Caldew tributary (southwest of the domain), possibly because the model fails to account for an observed build-up of debris under some bridges along this reach (Day, 2005). Also, we can see that the model predicts the flood

inundation extent marginally more accurately in the rural areas than the urban areas, and that the majority of the observations and flood edge are in rural areas. This is not unusual, and it may be desirable to use vulnerability-weighted calibration schemes in which the model is assessed at locations where flooding has the greatest consequence. Furthermore, maximum depth is only one component of flood hazard because other factors such as duration,

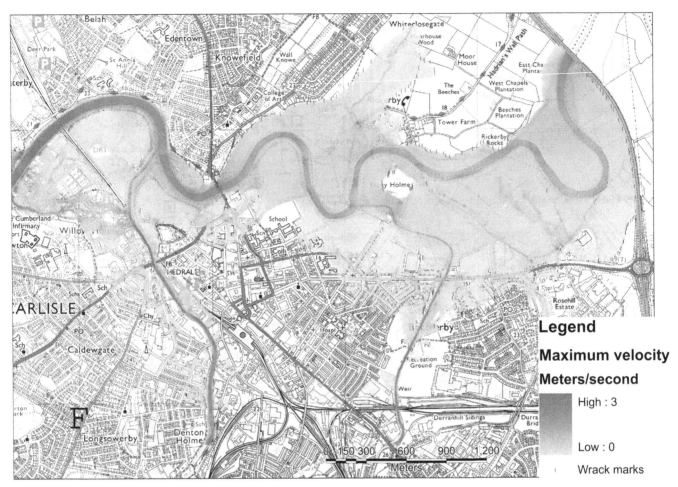

Figure 8.7 Map of maximum simulated velocity.

flow velocity and pollutant content can also be important. Fortunately the same dynamic 2D hydraulic modelling approaches can help to derive these outputs, particularly velocity, which is the volumetric flow rate between model cells divided by the cross-sectional area of flow. Flow velocities simulated by this model are shown in Figure 8.7. As expected, the greatest velocities are found in the channel, especially the lower sections of the River Eden, where the channel is bounded by embankments. High velocities of up to 3 m s^{-1} were simulated beneath the road bridge over the River Eden, which was expected given the morphology of the study site. Perhaps of more interest for hazard assessment are the maximum velocities in the urban areas of the domain, where we might assume the flood waters are most likely to come into contact with people and valuable structures. Velocities in these areas are generally in the range of 0.1–1 m s^{-1}, except in the Willow Holme area of the city, where flows from the Caldew were partially diverted through the urban area and away from the confluence with the Eden. This has potentially identified an area where hazard is significantly enhanced by the flow velocity to as much as 1.5 m s^{-1}. However, unlike the depth simulation, there are no validation data available to assess this aspect of the model behaviour.

8.4 DISCUSSION AND CONCLUSIONS

The above discussion clearly demonstrates that urban flood risk analysis requires data sets and models that are significantly more spatially resolved than those needed for rural studies. Typical rural flood simulation requires models and data with horizontal resolutions in the range 10–100 m (Horritt and Bates, 2001), yet Figure 8.5 makes clear that the appropriate resolution for urban hydraulic models is closer to 1 m. Topographic data at such resolutions are increasingly available through techniques such as airborne LiDAR survey, while opportunistic data sets to evaluate models from aerial photography and post-flood surveys of wrack and water marks using DGPS are increasing. It is therefore theoretically possible to build and evaluate urban flood inundation models at the native resolution of current airborne LiDAR data (~1 m) that should capture the overwhelming majority of feature length scales relevant to urban flood propagation. Nevertheless, such models represent a huge computational burden: for a typical shallow water model, where the minimum stable time step is determined by the Courant number, a halving of the grid resolution will result in an order of magnitude increase in the computational cost. Moving from 10 m to 1 m resolution models therefore incurs a ~1,000× increase in simulation times, and this is a severe impediment to hydraulic modelling at these scales. Fortunately, this problem is beginning to be addressed through the development of new equation sets (Bates *et al.*, 2010), parallelization (Neal *et al.*, 2009b, 2010), and the use of general purpose GPU (Lamb *et al.*, 2009) and HPC technologies.

Our ability to evaluate models is critically dependent on the quality of available model validation data; thus, we require more ways of mapping flood water elevations in urban areas. For rigorous model evaluation, improved means of characterizing and accounting for the uncertainty in our observations are needed in order to better define the plausible range of acceptable models. New methods for conducting rapid and accurate mapping of water levels during and after floods, greater availability of aerial photo data, and better methods to map urban flooding in new high-resolution satellite synthetic aperture radar sensors such as TerraSAR-X are all required. Moreover, we need to develop data sets that show the dynamic evolution of water levels through urban areas during flood events, to both improve our understanding of the dynamics of urban flood inundation and to better validate the dynamic performance of hydraulic models. Current data sets acquired by flood management and civil protection authorities tend to prioritize data capture at or around the flood peak in order to map maximum damage, yet such data may be relatively poor at discriminating between alternative model structures, particularly during valley-filling events where the shoreline lies on steep slopes at the edge of the floodplain (Bates *et al.*, 2004). Instead we require multiple synoptic maps of water surface elevation and extent at multiple times through flood events. One of the first studies of this nature was conducted by Neal *et al.* (2011) for the flood event that occurred in the UK town of Tewkesbury in summer 2007, and it yielded important insights into the resolution versus performance trade-offs in dynamic models of urban flooding.

Whilst there is much to be accomplished before we can claim to have a reliable and generic urban flood modelling capability, it is clear that considerable progress has been made in moving from rural to urban flooding inundation modelling in the last 5–10 years. This is beginning to yield new insights into the controls on urban flood inundation and the steps necessary for its accurate prediction.

9 Changes in flood propagation caused by human activities

This example application shows the use of 1D and 2D hydraulic models to simulate historical flood events and evaluate the effects on flood wave propagation of human activities, such as river training and levee heightening.

9.1 INTRODUCTION

The height of river levees (or dikes or embankments) has increased during the past two centuries and rivers have become increasingly controlled (Janssen and Jorissen, 1997). The heightening of levees to protect flood-prone areas results in increased damage if a failure occurs (Vis *et al.*, 2003). In fact, after the raising of levees, people feel safer and investments in the flood-prone areas increase. This is the so-called 'levee effect', caused by a false sense of safety where floodplain inhabitants perceive that all flood risk has been eliminated once a levee is raised (Burton and Cutter, 2008; Castellarin *et al.*, 2011). Given that risk can be defined as a combination of the probability of occurrence and its potential adverse consequences (Chapter 1), by heightening the levee systems the former (flooding probability) is reduced, but the latter (potential flood damage) might significantly increase. This clearly indicates that if proper socio-economic considerations are not seriously included in flood management, one might end up with the paradox that flood risk actually increases as a result of strengthening flood defence structures. Furthermore, with steadily increasing levee heights, the potential flood depth increases because the embankment of the river reduces the attenuation of floods. As an example, Figure 9.1 shows the geometry of a cross section of the River Po at Pontelagoscuro (Italy), which experienced a significant heightening of levees between 1878 and 2005. Figure 9.2 shows the increase in the length of the levee system of the River Po and its tributaries in the last two centuries and the corresponding increase in water depth observed at Pontelagoscuro during the largest historical floods.

The recent scientific literature has investigated the effects of human activity on flood wave propagation. For instance, Mitkova

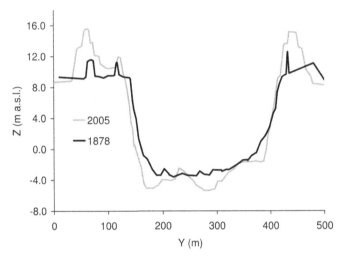

Figure 9.1 River Po at Pontelagoscuro (Italy): river cross section surveyed in 1878 and 2005 (Di Baldassarre *et al.*, 2009e).

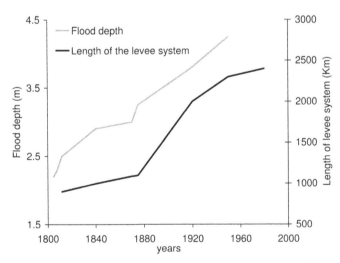

Figure 9.2 Evolution over time of the overall levee system for the River Po and its main tributaries in the last two centuries and corresponding increase of the maximum water depth at Pontelagoscuro observed during floods (Di Baldassarre *et al.*, 2009e).

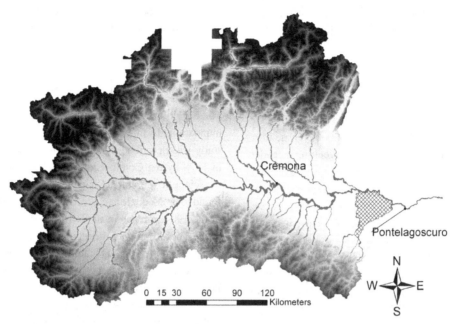

Figure 9.3 Test site: River Po basin (Northern Italy), river network (grey), location of Cremona and Pontelagoscuro, upstream and downstream limits of the model and location of the Secchia-Panaro region (River Po Basin Authority, www.adbpo.it).

et al. (2005) used a non-linear reservoir river model to analyse the flood propagation changes due to human activities in the Kienstock–Bratislava reach of the River Danube by simulating flood wave transformation for historical floods that occurred since 1899. Also, Natale *et al.* (2002) analysed the effects of river geometry modifications on a reach of the River Tiber (Italy) over a period of about 15 years by means of a mathematical model. In this context, this example application shows the use of 1D and 2D hydraulic models to simulate historical events and analyse the effects of human activities, such as the levee heightening on the River Po between 1878 and 2005 (Figure 9.1), on flood wave propagation in a 190-km reach of the middle–lower portion of the River Po.

9.2 TEST SITE AND PROBLEM STATEMENT

9.2.1 Case study

The River Po flows around 650 km eastward across Northern Italy, from the western Alps to the Adriatic Sea near Venice (Figure 9.3). The Po's drainage basin area is around 71,000 km², the largest Italian drainage basin area. The River Po is the longest river in Italy, and the largest river in terms of river discharge. The Po flows through many important Italian cities, including Turin and (indirectly) Milan, to which the Po is connected through a net of artificial and natural channels called Navigli, which was partly designed by Leonardo da Vinci (Di Baldassarre *et al.*, 2009e). Near the Po's mouth in the Adriatic Sea, the river creates a large delta, which consists of hundreds of minor channels and five main ones.

The example application was performed on a 190-km reach of the middle–lower portion of the River Po from Cremona to Pontelagoscuro (Figure 9.3). For this portion of the river, the bed slope is equal to about 0.2 m km⁻¹ and the riverbed consists of a stable main channel around 400 m wide and two lateral banks, characterized by an overall width between 200 m and 5 km, which is confined by two continuous artificial levees (Figure 9.1). We refer to two topographical ground surveys of the River Po: the first one performed by the Commissione Brioschi in 1878 (Di Baldassarre *et al.*, 2009e) and the second one performed by the Interregional Agency of the River Po (www.agenziainterregionalepo.it) in 2005 (Figure 9.1).

The River Po Basin Authority coordinates the management for the entire drainage basin area, which includes a number of different districts and the whole Po Plain (Pianura Padana), a very important agricultural region and industrial heart of Northern Italy. Figure 9.3 shows the so-called Secchia–Panaro region: the region bounded by the River Po and two tributaries, the Secchia and the Panaro. Three relevant flood inundations occurred in the nineteenth century in the Secchia–Panaro region: the first in November 1839, the second in October 1872, and the third in June 1879 (Govi and Turitto, 2000). In particular, during the end of May and the beginning of June 1879, the River Po experienced a significant flood event that caused the inundation of 432 km² in the Secchia–Panaro region (Figure 9.3). The inundation was due to a break in the right levee, which was caused by piping.

9.2.2   Problem statement

The riverbed of the Po has been modified by human interventions that began nearly two millennia ago and increased in frequency over time (Brath and Di Baldassarre, 2006). These interventions consisted of embankment construction, cutting meanders and implementation of excavations, which together have transformed the geometry of the river, so that nowadays the geometry could be considered artificial (Di Baldassarre et al., 2009e). The levee system of the River Po reached the current configuration as a result of a series of installation, development and consolidation over the centuries. These steps were strongly accelerated during the twentieth century. It is well known that the hydraulic works modified the natural expansion capacity of the river. Here a numerical test for clarifying these effects and for providing a plausible quantification is carried out as an example application. More specifically, a series of topographic, hydrologic and inundation data referred to the June 1879 flood event (inundated area equal to 432 km^2, water depth in flood-prone areas equal to 6 m) were collected. Then, flood inundation models were used to: (i) reconstruct a historical flood inundation that occurred in 1879 and (ii) assess the effects of human activities on flood wave propagation in the middle–lower portion of the River Po.

9.3   METHODS

9.3.1   Hydraulic modelling

Numerical simulations were performed by using a hybrid methodology. In particular, a 1D approach was used to simulate flows in the river and in the unprotected floodplain (i.e. within the levee system). Flows through the levee breach were also computed by the 1D code (HEC-RAS, Hydrologic Engineering Center, 2001) and then adopted as the inflow boundary condition for a 2D model (TELEMAC-2D, Galland et al., 1991) of the flood-prone area, protected by the levee system. Hence, the two models are not linked interactively. Given that the Po levees are about 10 m higher than the flood-prone area, the assumption that there are no interactions between the two models was found to be reasonable. In particular, Aureli et al. (2006) demonstrated, for a similar reach of the River Po, that this type of modelling is appropriate. In particular, they compared a fully 2D model to a 1D–2D model in reproducing inundation scenarios due to levee failure (Aureli et al., 2006). The study pointed out that the two models have a similar performance. Nevertheless, the 1D–2D approach has the advantage that the dynamic flooding could be predicted whilst

avoiding the onerous description of the riverbed geometry in 2D and, consequently, achieving a reduction in the computational time.

9.3.2   Model evaluation: the 1879 inundation

The hybrid approach was used to simulate the June 1879 flood inundation. To this end, the topography of the river reach under study was described by using historical cross sections surveyed in 1878 by the Commissione Brioschi. Based on indications in the scientific literature (Chapter 4), the cross-section spacing was found to be adequate to correctly describe the hydraulic behaviour of the 190-km reach of the River Po during the June 1879 flood event. The altimetry of the region was determined by means of the 100-m resolution DTM (Figure 9.3), assuming that the variations which occurred in the last century on the elevation of the protected flood-prone areas are negligible.

The model used the observed flow hydrograph at the upstream end of the reach (Cremona) derived by Galloni (1881) as the upstream boundary condition, and a rating curve at the downstream end of the reach (Pontelagoscuro). The friction coefficients were differentiated for main channel and floodplain and estimated according to the results of previous studies (Aureli et al., 2006; Brath and Di Baldassarre, 2006; Di Baldassarre et al., 2009e) and indications reported in the scientific literature (see e.g. Chow, 1959) on the basis of the physical characteristics of the river (0.04 m$^{-1/3}$ s for the main channel and 0.09 m$^{-1/3}$s for the floodplain). The levee breach was analysed within HEC-RAS by modelling the levee as a lateral structure (Barkau, 1997). The characteristics of the breach were imposed by using the historical observations (Galloni, 1881): the levee failure was caused by piping, started on 4 June 1879 at 4 a.m., and was characterized by a final width of 220 m. The breach was assumed to occur instantaneously as there was no information about the evolution in time of the breach. However, a preliminary sensitivity analysis on the time of formation of the breach pointed out that the assumption of instantaneous breach does not influence the results of the model (Di Baldassarre et al., 2009c). The model was verified by comparing the simulated water elevations with an observed stage hydrograph. Figure 9.4 shows the simulated and the observed water levels at Ostiglia (Galloni, 1881), which is located in a central position along the considered reach. The time of the levee breach, which occurred after the peak flow, is clearly visible. Also, Figure 9.4 shows the good agreement between simulated and observed water elevations.

The 2D model was then used for simulating the flow in the 720-km^2 flood-prone area. The computational mesh was characterized by 31,702 elements and 17,605 nodes. The friction coefficient of the flood-prone area was assumed to be constant and equal to the value for the unprotected floodplain (0.09 m$^{-1/3}$ s). The 2D model results were then compared to the historical flood extent

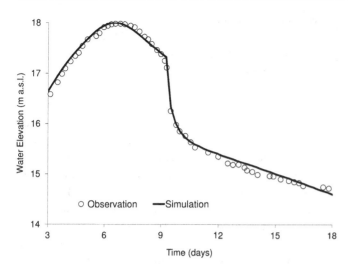

Figure 9.4 One-dimensional model results: water surface elevation in the middle of the reach (Ostiglia) during the June 1879 event; measured values (Galloni, 1881) and model results.

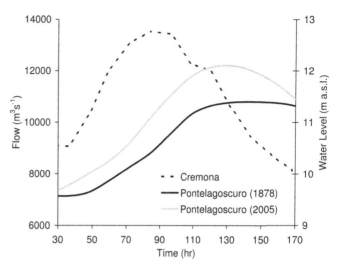

Figure 9.5 The effects of levee heightening and river geometry modification on flood wave propagation: upstream boundary condition at Cremona (flow hydrograph), and the resulting water levels downstream at the Pontelagoscuro station using the 1878 geometry and the 2005 geometry (Di Baldassarre *et al.*, 2009e).

map of the June 1879 event. The extent of agreement between the historical flood extent map and the model simulations was approximately equal to 90% (Di Baldassarre *et al.*, 2009e). Given that the total inundation extent is confined by the topography, this result provides mainly an indication about the correct estimation of the total outflow volume (around $1,200 \times 10^6$ m^3). Moreover, the flood inundation model performance was tested by the historical information concerning the 1879 flooding dynamic (e.g. Galloni, 1881; Govi and Turitto, 2000), such as the averaged front wave velocity (around 15 km/day, historical observation; 13.5 km/day, simulation) and time to reach Bondeno town (around 40 hours, historical observation; 42 hours, simulation). The performance of the 2D model in reproducing flood wave propagation over an initially dry plane indicates that this type of modelling can be used as a tool for the implementation of proper civil protection plans (e.g. evacuation plans).

9.3.3 Numerical experiment

The 1878 and 2005 geometries (Figure 9.1) were used to analyse the effects of the development and consolidation of the levee system of the River Po on flood wave propagation characteristics. To this end, two different models were built: the first, indicated as the 1878 model, using the river geometry surveyed in 1878; the second, indicated as the 2005 model, using the river geometry surveyed in 2005. To make a fair comparison, this study was performed as follows: (i) the entire levee system was assumed to be non-erodible (no formation of breach when the levee is overtopped); (ii) even though the 2005 survey was characterized by a higher number of cross sections, the 2005 model utilized the same number of cross sections (i.e. the same cross-section spacing) as

the 1878 model; (iii) both models used the same Manning's coefficients (see Section 9.4). Assuming the same roughness values for both models (1878 and 2005) is consistent with the goals of the analysis. In fact, the main objective here is to assess the changes in flood propagation exclusively due to levee heightening and river geometry modifications, without introducing additional differences between the two models.

While the 1878 model was evaluated by reproducing the 1879 inundation (Section 9.3.2), the 2005 model was evaluated by referring to the recent flood event of October 2000 and comparing the model results to the high water marks surveyed in the aftermath of the flood as well as to observed stage hydrographs. This evaluation showed the good performance of the model, as the mean absolute error between observed and simulated water levels was found to be equal to 0.3 m (1–2% of the water depth).

The numerical experiment was carried out by using, as hydrologic input, the flood hydrograph observed at the upstream end of the reach (Cremona) during the catastrophic November 1951 flood (Figure 9.5), the most important flood of the last century. The 1951 flood impacted several structures during its passage, and an extensive inundation occurred in the lower portion of the river. In particular, the left levee next to Occhiobello was overtopped and breached. The total volume of the flow leaving was equal to around $2,000 \times 10^6$ m^3; it produced an inundation extent equal to around 1,100 km^2 and some 380,000 people were evacuated.

The 1951 flood hydrograph has the meaning of a synthetic event for the two geometries. In fact, due to the different levee system along tributaries in 1878, the 1951 flood event would

Table 9.1 *Results of the numerical analysis on the effects of the levee heightening on flood propagation*

Geometry	Peak flow at Pontelagoscuro	Outflow volume	Inundation extent in protected areas
1879	10,500 m^3 s^{-1}	450 × 10^6 m^3	200 km^2
2005	12,000 m^3 s^{-1}	--	--

have resulted in a more attenuated flood wave at Cremona than the one illustrated in Figure 9.5. However, the adoption of an overestimated flood wave at Cremona is functional to the goal of this simulation. In fact, it allows an understanding of the effect of levee heightening and geometry modification, which occurred in the last century, on the hydraulic behaviour of the low–middle portion of the River Po during extreme floods.

9.4 RESULTS

The analysis pointed out some interesting results for the quantification of the loss of expansion capacity (i.e. flood attenuation) between 1878 and 2005. Figure 9.5 reports the upstream hydrograph used for both models (i.e. hydrograph recorded in 1951 at Cremona) and the downstream hydrograph at Pontelagoscuro obtained with the two different models (1878 and 2005). Figure 9.5 shows that the simulated water levels downstream (at Pontelagoscuro) obtained with the 1878 model are more attenuated than the ones obtained with the 2005 model. Table 9.1 summarizes the results of this numerical analysis. In particular, the peak flow at Pontelagoscuro with the 1878 geometry is around 15% lower than the value obtained with the 2005 geometry (Table 9.1). The different attenuation of the flood wave is mainly caused by the presence of flooding in prone areas along the reach due to overtopping of the 1878 levee system, which would have not been able to contain the 1951 flood event.

As expected, the levee heightening of the River Po reach analysed here had two contrasting effects. On the one hand, it had a positive effect as it decreased the frequency of inundation of the flood-prone area: during the nineteenth century in the Secchia–Panaro region three historical inundations were registered, while during the twentieth century there were no inundations in such flood-prone areas. On the other hand, it decreased the natural expansion capacity of the River Po reach analysed here, consequently increasing flood discharge downstream (Figure 9.4 and Table 9.1). Moreover, the numerical analysis allowed the quantification of the effects of levee heightening. The most important result is that, while during the 1951 event the absence of flooding along the reach (mainly due to levee heightening in the period 1878–1951) resulted in a higher flood discharge downstream and

consequently produced the disastrous levee failure downstream (Table 9.1), with the 1878 geometry the 1951 flood event would have produced inundation in flood-prone areas for a modest total amount of water (450 × 10^6 m^3 instead of 2,000 × 10^6 m^3).

This result gives useful indications for the choice of the most appropriate strategy for future flood risk management. In fact, although with the 2005 geometry the River Po is able to contain the entire 1951 flood event without any levee overtopping and inundation of flood-prone areas (Table 9.1), disastrous levee failures may be expected for events of a higher magnitude (e.g. 500-year flood event; Maione *et al.*, 2003). Therefore, in the case of higher magnitude flood events, in order to avoid catastrophic inundation downstream (as during the 1951 event), the results of this numerical analysis recommend the use of alternative approaches, such as controlled flooding of certain areas (e.g. Vis *et al.*, 2003; European Parliament, 2007; Komma and Blöschl, 2008), instead of continuous levee heightening. In such an approach, flooding may be allowed in certain areas (identified on the basis of an accurate cost–benefit analysis), whilst the impact of flooding is minimized through policies of appropriate land-use planning (Chapter 11).

9.5 CONCLUSIONS

This application example aimed to assess the applicability of 1D–2D hydraulic models for reconstructing a historical flood inundation and to evaluate the impacts of human modifications on flood wave propagation.

The example showed the applicability of the flood inundation model to reconstruct inundated areas and flooding dynamics of the June 1879 inundation. Moreover, the results of the study allowed the assessment of how the levee heightening changed the flood wave propagation characteristic for the river reach analysed. It is well known that levee heightening has a twofold effect: on the one hand, it decreases the frequency of inundation of flood-prone areas; on the other hand, it decreases the flood attenuation, consequently increasing the flood discharge downstream. This study quantified such effects through an analytical comparison of the hydraulic answer to the same hydrologic input, the November 1951 flood.

This example is a first attempt to investigate the effects of human interventions on flood propagation. However, the results of this study give some useful indications for planning future strategies in flood risk management. In particular, they suggest that alternative strategies based on the promotion of sustainable land-use practices, improvement of water retention as well as controlled flooding in certain areas (Chapter 11) are often to be preferred to the more common and traditional approach based on continuous embankment and levee heightening.

Table 9.2 *Time series of the annual maximum values of river discharge at the given cross section*

Year	Discharge (m³ s⁻¹)	Year	Discharge (m³ s⁻¹)
1975	56	1992	44
1976	48	1993	35
1977	40	1994	52
1978	35	1995	40
1979	12	1996	61
1980	34	1997	37
1981	40	1998	78
1982	40	1999	27
1983	56	2000	55
1984	34	2001	27
1985	28	2002	74
1986	52	2003	35
1987	84	2004	35
1988	32	2005	49
1989	34	2006	21
1990	27	2007	82
1991	88	2008	27

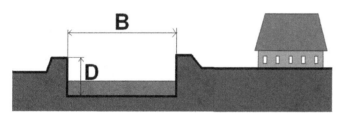

Figure 9.6 Scheme of the exercises. Note that the scale of the diagram is not realistic as it is only meant to illustrate the exercises.

9.6 EXERCISES

The exercises of this chapter deal with the aforementioned concept of 'levee effect'. To this end, Figure 9.6 shows a river characterized by a large rectangular cross section ($B = 60$ m; $D = 2.5$ m) and a constant bed slope ($S_0 = 0.0001$). Table 9.2 reports the annual maximum values of the river discharge (m³ s⁻¹) observed at this given cross section.

Specific tasks

9.1. Derive the 1-in-100 year flood, Q_{100} (i.e. design flood with a return period of 100 years), by using the Gumbel distribution and the method of moments.

Useful equations

The Gumbel distribution (or extreme value 1; see e.g. Chow *et al.*, 1998) is widely used in applied hydrology and hydraulic

engineering for the inference of extreme values (e.g. flood data). Its main advantage is in being a parsimonious model (two parameters only; θ_1 and θ_2) that typically fits extreme values reasonably well (e.g. Laio *et al.*, 2009). The Gumbel cumulative distribution function of the distribution can be defined as follows:

$$F_X(x) = \exp\left\{-\exp\left[-\frac{x - \theta_1}{\theta_2}\right]\right\} \quad (9.1)$$

From equation (9.1), the quantile can be easily derived:

$$x_F = \theta_1 - \theta_2 \ln\left[-\ln\left(F\right)\right] \quad (9.2)$$

The literature has presented many techniques for the estimation of the Gumbel parameters, θ_1 and θ_2 (e.g. moments, maximum likelihood, L-moments; see Laio *et al.*, 2009). For the sake of simplicity, this exercise can be carried out by using the method of moments where θ_1 and θ_2 can be explicitly derived from the sample mean (μ) and standard deviation (σ):

$$\begin{aligned} \theta_2 &= 0.78\sigma \\ \theta_1 &= \mu - 0.57772\theta_2 \end{aligned} \quad (9.3)$$

Hence, under the assumption that the annual maximum values of the river discharge (Table 9.2) are a time series of a random variable, θ_1 and θ_2 can be estimated for this exercise by simply computing the mean (μ) and standard deviation (σ) of the values reported in Table 9.2 and applying equations (9.3).

Then, by considering equation (9.2) and the definition of the return period in years (T) as the average recurrence interval, the flood quantile Q_T (1-in-T year flood event) can be estimated as

$$Q_T = \theta_1 - \theta_2 \ln\left(-\ln\left(1 - \frac{1}{T}\right)\right) \quad (9.4)$$

More details can be found in many textbooks of applied hydrology, such as Chow *et al.* (1998).

9.2. Evaluate the water level corresponding to the 1-in-100 year flood, under the assumption of uniform flow and roughness coefficient (n) equal to 0.035m⁻¹/³ s.

Useful equations

For this exercise, to evaluate the water level, h_{100}, corresponding to the 1-in-100 year flood under the assumption of uniform flow and large rectangular section, equation (2.4) can be rewritten as

$$Q_{100} = \frac{1}{n} B h_{100}^{5/3} S_0^{1/2} \quad (9.5)$$

Thus, the water level corresponding to the design flood can be easily written:

$$h_{100} = \left(\frac{n Q_{100}}{B S_0^{1/2}}\right)^{3/5} \quad (9.6)$$

More details can be found in many textbooks of open channel hydraulics, such as Chow (1959).

9.3. Is the cross section ($D = 2.5$ m) able to convey the 1-in-100 year flood?

9.4. Evaluate the needed heightening of the levee system design, without changing the width of the cross section (Figure 9.6), to convey the 1-in-100 year flood.

9.5. Reflect on the fact that levee heightening may lead to the development of flood-prone areas (Figure 9.6) because people feel safer. Thus, by raising levees, on the one hand the probability of flooding decreases, on the other hand the potential adverse consequences of floods increase ('levee effect').

10 Changes of stage–discharge rating curves

This example application shows a hydraulic study of the historical changes of stage–discharge rating curves caused by changes in river geometry. The application also points out the utility of hydraulic models to extrapolate rating curves beyond the measurement range.

10.1 INTRODUCTION

River flow data are usually affected by a relevant uncertainty, especially when the stage–discharge rating curve is extrapolated beyond the measurement range used for its derivation (Chapter 4). Hydraulic models have been shown to be useful tools to reduce the inaccuracies due to the extrapolation of rating curves (Horritt *et al.*, 2010). This chapter shows an example application where the changes of stage–discharge relationships, caused by modifications of the river geometry, are hydraulically analysed. In particular, five hydraulic models of a 16-km reach of the River Po (Italy) were built using five topographical ground surveys (performed in 1954, 1968, 1979, 1991 and 2000) as geometrical inputs. These five models were then used to investigate and discuss the hydraulic behaviour of the river reach and, in particular, to assess the effects of river geometry changes in the stage–discharge rating curves.

10.2 TEST SITE AND PROBLEM STATEMENT

The example application refers to a 16-km reach of the River Po (Northern Italy), between Cogozzo and Tagliata. Five different topographical ground surveys of this river reach, performed in 1954, 1968, 1979, 1991 and 2000, were used in this study. As an example, Figure 10.1 shows the historical surveys of an internal cross section (Viadana).

The aforementioned surveys were then used as geometrical inputs of five different HEC-RAS models (Hydrologic Engineering Center, 2001). River discharge at the upstream end and friction slope at the downstream end define the model boundary

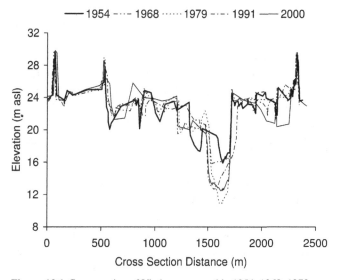

Figure 10.1 Cross section of Viadana surveyed in 1954, 1968, 1979, 1991 and 2000.

conditions. Concerning the roughness parameters, in order to avoid subjectivity in separating the main channel from the floodplain for each cross section in the five different topographical ground surveys, a uniform Manning's coefficient for the entire cross section (channel and floodplain) is utilized. This assumption is justified by the findings of previous studies performed in the same river reach using HEC-RAS (Di Baldassarre *et al.*, 2009e). In particular, Di Baldassarre *et al.* (2009e) calibrated a HEC-RAS model using a large amount of data on the October 2000 flood event. The calibration exercise showed that the optimal set of parameters agrees well with the values given in standard tables of Manning's coefficients (0.04 m$^{-1/3}$ s for the channel and 0.09 m$^{-1/3}$ s for the floodplain; Chow, 1959). The same study demonstrated that parameter compensation, due to Manning's coefficient decrease in the floodplain and its increase in the main channel, allows one to use a uniform Manning's coefficient for the whole section, equal to around 0.05 m$^{-1/3}$ s, while preserving almost equivalent performance of the hydraulic model (Di Baldassarre and Claps, 2011).

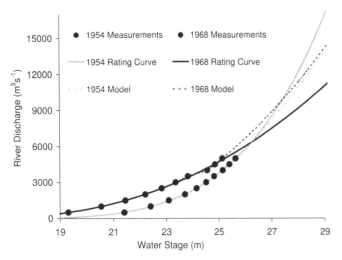

Figure 10.2 Results of the first experiment in Viadana: construction of the rating curves using two different geometries (1954 and 1968). The figure also shows the results of the hydraulic model (dotted lines) (Di Baldassarre and Claps, 2011).

10.3 METHODS

This study focuses on the extrapolation errors of the steady rating curve and is made by means of numerical experiments. Hence, it is important to note that in steady flow conditions, for this river reach it is reasonable to assume the presence of a one-to-one correspondence between the water stage and the river discharge, because of the minor role played by downstream disturbances and tributaries (Franchini *et al.*, 1999; Di Baldassarre and Montanari, 2009; Di Baldassarre and Claps, 2011).

The first numerical experiment was carried out to investigate how the river geometry modification affects steady-state rating curves. The experiment focuses on an internal cross section (Viadana, Figure 10.1) and uses two different geometries surveyed in 1954 and 1968. Specifically, steady-state simulations with the hydraulic model produce 'measured' river discharges values. These are in the range between 500 m^3 s^{-1} (low flow condition) and 5,000 m^3 s^{-1} (ordinary flood condition), in steps of 500 m^3 s^{-1}. The rating curve expressed by equation (4.5) is estimated by interpolating the (Q, h) points. These simulations are run using a uniform Manning's coefficient equal to 0.05 m$^{-1/3}$ s and the least squares method is used to estimate the three parameters of the power-law equation (Chapter 4; equation 4.5). The choice of the discharge interval reflects actual practice in making direct measurements of river discharge up to ordinary flow conditions (Franchini *et al.*, 1999). This is obviously due to the fact that measuring discharge during extreme floods is very difficult, if not impossible.

Figure 10.2 shows the results of this first numerical experiment and clearly demonstrates that the two rating curves are strongly different. Differences in the interpolation zone (500–

5,000 m^3 s^{-1}) reflect the changes in the natural geometry of the River Po that occurred in the period 1954–68 (Figure 10.1). In contrast, the high differences in the extrapolation zone (5,000–12,000 m^3 s^{-1}) cannot be justified by data and just reflect the curves' shape in the extrapolation range. More specifically, Figure 10.2 shows that the water stage of 30 m a.s.l., which is the elevation of the levee system (Figure 10.1), would correspond to around 15,500 m^3 s^{-1}, according to the 1954 rating curve, or around 12,000 m^3 s^{-1}, according to the 1968 rating curve. This difference appears too large: it is hard to believe that the river geometry modification in the period 1954–68 would have led to a decrease of the hydraulic capacity of the river reach from 15,500 m^3 s^{-1} to 12,000 m^3 s^{-1}. More details about this experiment are described in Di Baldassarre and Claps (2011).

To better investigate the rating curve behaviour in the flood discharge range, a second set of numerical experiments was carried out, reconstructing water surface profiles by using the five topographical ground surveys as geometric input. The hydraulic simulations were performed by imposing river discharge from 500 m^3 s^{-1} to 12,000 m^3 s^{-1}, in steps of 500 m^3 s^{-1}, and, again, a uniform Manning's coefficient equal to 0.05 m$^{-1/3}$ s.

Table 10.1 reports the outcomes of these simulations in terms of water stage corresponding to a given river discharge at Viadana. Differences in the water stage (for a given discharge) reported in Table 10.1 are caused by the changes in the cross-section geometry, including the cease-to-flow stage. The last column of Table 10.1 reports the standard deviation of the water stage values. It is interesting to note that, although considerable changes occurred in the geometry of this river reach (Figure 10.1), the flood stages corresponding to high discharge values remain approximately constant (Table 10.1). More specifically, standard deviations of the water stage remain in the range of 20–30 cm, when the discharge exceeds 5,000 m^3 s^{-1} (Table 10.1). It is important to underline that 20–30 cm represents the tolerance of results of computational hydraulic models in view of the many other sources of inaccuracy (Di Baldassarre and Claps, 2011). Moreover, Figure 10.2 compares the hydraulic model results with the rating curves derived using the analytical relationship (Chapter 4, equation 4.5). It is interesting to note that for high-flow conditions, while the hydraulic results tend to converge, the two rating curves diverge.

A third set of experiments was performed to take into account the uncertainty of the model parameters (Chapter 6). For each geometry (1954, 1968, 1979, 1991 and 2000), the numerical computations were run using three values of the Manning's coefficient, i.e. 0.045 m$^{-1/3}$ s, 0.050 m$^{-1/3}$ s, 0.055 m$^{-1/3}$ s. Figure 10.3 shows the results obtained for the internal cross section at Viadana in terms of standard deviation of the water stage versus river discharge for different Manning's coefficients. This third experiment confirmed the results of the second experiment (Table 10.1): despite considerable modifications occurring in the geometry of

Table 10.1 *Results of the second experiment: simulated water stage (m) at Viadana cross section versus river flow values (m^3 s^{-1}). The last column reports the standard deviation (m) of the water stage.*

Q	1954	1968	1979	1991	2000	Standard deviation
500	21.55	19.48	17.92	18.28	18.20	1.50
1,000	22.57	20.77	20.05	20.14	20.08	1.07
1,500	23.33	21.72	21.24	21.30	21.34	0.88
2,000	23.93	22.50	22.18	22.18	22.32	0.74
2,500	24.35	23.12	22.91	22.95	23.12	0.60
3,000	24.72	23.67	23.49	23.55	23.71	0.51
3,500	25.08	24.49	24.00	24.09	24.23	0.43
4,000	25.38	24.83	24.40	24.50	24.67	0.39
4,500	25.64	25.15	24.75	24.86	25.07	0.34
5,000	25.90	25.45	25.07	25.19	25.44	0.32
5,500	26.14	25.74	25.38	25.50	25.77	0.29
6,000	26.38	26.01	25.67	25.79	26.09	0.28
6,500	26.61	26.27	25.95	26.07	26.46	0.27
7,000	26.83	26.53	26.21	26.34	26.73	0.26
7,500	27.05	26.77	26.47	26.59	26.99	0.25
8,000	27.26	27.01	26.71	26.84	27.23	0.24
8,500	27.47	27.25	26.95	27.08	27.47	0.23
9,000	27.67	27.47	27.18	27.31	27.71	0.23
9,500	27.87	27.69	27.41	27.53	27.93	0.22
10,000	28.07	27.90	27.63	27.75	28.15	0.22
10,500	28.27	28.11	27.84	27.96	28.37	0.22
11,000	28.46	28.32	28.05	28.17	28.58	0.21
11,500	28.65	28.52	28.26	28.37	28.78	0.21
12,000	28.83	28.72	28.46	28.57	28.99	0.21

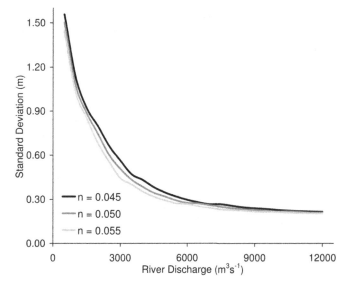

Figure 10.3 Results of the third experiment: standard deviation of the water stage versus river discharge for three different Manning's coefficient values at the Viadana cross section.

the main channel of the River Po (Figure 10.1), the water stage corresponding to river discharge values higher than 5,000 m^3 s^{-1} appears independent of the specific river geometry as the standard deviation tends to 20–30 cm (independent of the river roughness; Figure 10.3). This outcome has a physical explanation: changes in the geometry of the river reach under study have largely occurred in the main channel (Figure 10.1) and therefore they have a minor effect on the hydraulics of floods where the floodplain gives a relevant contribution to the flow. This hypothesis, although appropriate for many alluvial rivers, is not applicable as a general rule. For instance, if the floodplain width is not much larger than the channel width, changes in floodplain geometry due to sediment deposition cannot be neglected (e.g. Swanson *et al.*, 2008). Moreover, human interventions (navigation, excavation) may produce significant alterations in the floodplain geometry. However, for the river reach under study, which has a bankfull discharge of about 3,000 m^3 s^{-1}, discharge values higher than 4,000–5,000 m^3 s^{-1} are representative of flow conditions in which floodplains provide a significant contribution to the flow and the stage–discharge relationships tend to be similar (Figure 10.3). Thus, these last two numerical experiments corroborate that differences found in the extrapolation zone of the rating curves (Figure 10.2) find little justification on changes in river geometry. To corroborate these findings, Di Baldassarre and Claps (2011) analysed the depth–width curves for the River Po cross sections, and pointed out that for high values of the water depth the top width tends to converge to a certain value. This represents a reasonable explanation of the fact that the stage–discharge relationships tend to be similar for high-flow conditions. It is important to note that this type of depth–width curves is typical of many alluvial rivers where the flood shoreline is constrained by slopes (or defences) bounding the floodplain, and therefore large changes in water depths produce small changes in lateral flood extent (Hunter *et al.*, 2007).

10.4 RESULTS

The results of this application example confirm that the indirect observation of discharges beyond the measurement range should rely on physically based models (e.g. HEC-RAS), instead of traditional approaches of extrapolating rating curves based on analytical relationships (Chapter 4, equation 4.5). A hydraulic study of the river reach is made possible nowadays by the broad availability of topographic data and model codes, and it may help to reduce the uncertainty in derivation of river discharge measurements, leading also to more reliable stage–discharge relationships in the extrapolation zone. A good operational strategy could be to use the stage–discharge measurements to calibrate a hydraulic model and then to use the model to extrapolate the rating curve. A hydraulic approach can also potentially include roughness variations due to changes in the state of the vegetation, which can be a

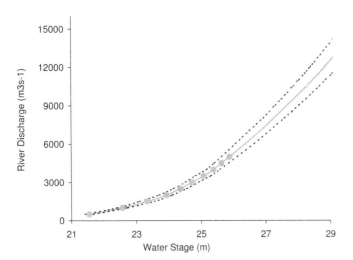

Figure 10.4 Example of hydraulically derived rating curve with uncertainty bounds where the measurements (grey dots) are used as calibration data (Di Baldassarre and Claps, 2011).

relevant factor in alteration of the rating curve (Di Baldassarre and Montanari, 2009). However, it must be said that the uncertainty of the hydraulic model, which is calibrated using ordinary flow data and then used to simulate extremely high flow conditions, cannot be neglected (Jarret, 1987; Kirby, 1987; Burnham and Davis, 1990). For instance, a number of studies (e.g. Horritt and Bates, 2002; Romanowicz and Beven, 2003; Horritt *et al.*, 2007) have shown that the effective roughness coefficients may be different when evaluated for different flow conditions (see also Chapter 6). It is then recommended to complement the model by associating the estimation of model uncertainty.

A rigorous and statistically consistent analysis of the uncertainty of the hydraulically derived rating curve is not an easy task and might be computationally infeasible. Hence, this chapter reports, as an example, a simple and pragmatic approach based on the widely used generalized likelihood uncertainty estimation (GLUE) (Beven and Binley, 1992; Pappenberger *et al.*, 2006; Montanari, 2007; and Chapter 6). In this approach the uncertainty of the hydraulic model is estimated as follows: (i) the hydraulic model is run using uniformly distributed roughness coefficients (selected according to prior knowledge) in the range 0.04–0.07 m$^{-1/3}$ s; (ii) the simulation results are compared to the

calibration data (i.e. stage–discharge measurements) and simulations with a mean absolute relative error higher than 20% are rejected as non-behavioural; (iii) the computed likelihoods are rescaled to produce a cumulative sum of 1, and then uncertainty bounds and the median simulation are derived by following the standard GLUE methodology (e.g. Montanari, 2005).

Figure 10.4 shows the hydraulically derived rating curve and the corresponding uncertainty bounds. It is important to note that the uncertainty bounds derived within the GLUE framework are unavoidably affected by a number of subjective decisions (see e.g. point ii) and reflect only the uncertainties in the model parameters, disregarding many other sources of uncertainty (Chapter 6).

10.5 CONCLUSIONS

Flood risk management studies and several hydrologic applications require the use of discharge data referred to flood conditions. However, several studies pointed out that the higher the flow, the higher the uncertainty of the rating curves that for these flow conditions are used far beyond the actual discharge measurements range. This example application showed that analytical functions commonly used to interpolate river discharge measurements (e.g. Chapter 4; equation 4.5) fail to reproduce the stage–discharge relationship in the extrapolation zone and can lead to results that are not physically plausible. Hence, a hydraulic approach to derive stage–discharge curves (with uncertainty) is recommended (see also Horritt *et al.*, 2010).

Another interesting result of this example application is that, for river discharge values sufficiently higher than the bankfull discharge, differences in water stage due to changes of river geometry tend to vanish. This is because changes in the geometry of the river reach mainly occur in the main channel and therefore do not have a strong effect on the hydraulics when the floodplain gives a relevant contribution to the flow. As previously discussed, although the geomorphological features of the River Po can be considered representative for many alluvial rivers in Europe and around the world, the results of this study should be further expanded in light of additional studies relative to different test sites.

11 Evaluation of floodplain management strategies

This example application shows the utility of 1D and 2D hydraulic models to evaluate and compare floodplain management policies. To this end, a methodology to produce flood hazard maps in areas protected by river embankments is described. The methodology is able to deal with uncertain localization, geometry and development of levee breaches.

11.1 INTRODUCTION

During the last two centuries, in many more-developed countries, rivers have become more and more controlled and the height of river dikes (i.e. levees or embankments) has increased (Janssen and Jorissen, 1997). However, it has been shown that with steadily increasing embankment heights the potential flood depth increases (Chapter 9). Also, levee heightening, which aims at protecting flood-prone areas, tends to increase the potential flood damage because of the aforementioned 'levee effect' (see Chapter 9).

Flood risk management strategies based on the construction, heightening and strengthening of embankments can be called resistance strategies (Vis *et al.*, 2003). In this approach, the design of river embankments and other water-retaining structures is usually based on an acceptable probability of overtopping; while the portion of risk that remains is called residual risk (van Manen and Brinkhuis, 2005). Residual flood risk behind levees is often not taken into account. In particular, given that levees are usually characterized by a uniform safety level (e.g. return period equal to 200 years), river discharges above the design flood (e.g. 1-in-200 year flood) might cause flooding anywhere and even at several locations at the same time, and therefore the evolution of the flood event is unpredictable. It is obvious that this condition is undesirable. For instance, in the case of exceptional events a large area must be evacuated, as all areas theoretically have the same probability of flooding.

A different approach to flood risk management is the so-called resilience strategy. The concept of resilience originates from ecology (e.g. Holling, 1973) and was introduced, in the context of flood risk management, by De Bruijn and Klijn (2001). The idea behind the resilience approach is 'living with floods' instead of 'fighting floods'. In this approach, flooding can be allowed in certain areas, and the impact of flooding is minimized through policies of land-use planning and management (Vis *et al.*, 2003).

Directive 2007/60/Ec of the European Parliament (2007) states that flood risk management policies may comprise the promotion of sustainable land-use practices, improvement of water retention, as well as controlled flooding of certain areas in the case of extreme flood events. As a result, many river authorities are currently evaluating the opportunity to implement alternative flood mitigation measures, such as controlled flooding, instead of continuous levee heightening and strengthening (see also Chapter 9). Flood hazard mapping, based on hydraulic modelling, may assist this process and enable the comparison of alternative strategies for flood risk mitigation and management (Di Baldassarre *et al.*, 2009c).

In this context, this example application shows the utility of hydraulic models to evaluate floodplain management policies. More specifically, probability-weighted hazard maps are used to compare two different flood protection strategies in an Italian test site: a traditional resistance strategy based on the use of a regular levee system, and an alternative approach based on the use of a hydraulic structure that allows controlled flooding of certain areas, where the expected flood damage is limited.

This chapter also describes an innovative approach for producing probability-weighted hazard maps (Di Baldassarre *et al.*, 2009c), based on an ensemble of numerical simulations. In this approach, several inundation scenarios (corresponding to different levee breach locations, geometries and evolutions in time) are simulated by coupling 1D and 2D flood inundation models. Then, the results of each scenario are combined to build inundation hazard maps and compare different flood mitigation strategies.

11.2 TEST SITE AND PROBLEM STATEMENT

11.2.1 Case study

The example application is performed on a 270-km^2 flood-prone area protected by the left embankment of a 28-km reach of

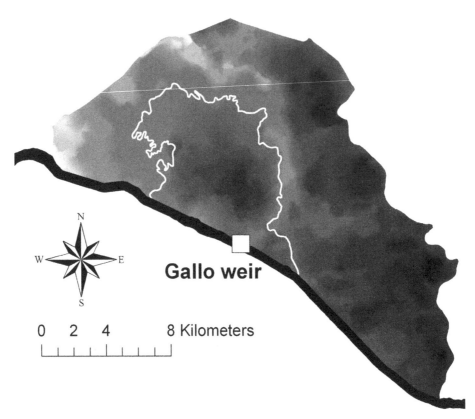

Figure 11.1 Test site: 28-km reach of the River Reno (Italy) considered in the study (black line, the river flows from west to east); 10-m resolution DTM (greyscale from 3 m a.s.l., white, to 15 m a.s.l., black); contour of the area inundated during the 1949 flood event (white line); and location of the Gallo weir (white square).

the River Reno (Northern Central Italy). The entire study area (Figure 11.1) is bounded by the embankments of the national road SS 255 and the Bologna–Venezia railway, the left levees of the River Reno, and the right levees of the River Po Morto di Primaro. The study area consists of agricultural land, residential areas and a few industrial plants.

This test site has been affected by two major inundations: the first in 1949 (Figure 11.1), which inundated an area equal to around 60 km², and the second in 1951, which inundated an area equal to around 116 km². Both inundation events were caused by breaches of the left embankment. The topographic data available for the study are: a 10-m resolution DTM (Figure 11.1); a survey of the principal breaklines (levees, road embankments, railway embankments); and 35 cross sections of the 28-km reach of the River Reno.

Over recent decades, regulation works have changed the geometry of the river. However, the flood exposure of this area is very high: about 25,000 inhabitants and a number of industrial sites are protected by extremely high (8–11 metres) river embankments and the riverbed is hanging, since its altitude is higher than the flood-prone area.

11.2.2 Problem statement

After the inundation of 1951, a lateral weir, 100 m wide, called the Gallo weir, was constructed in the left embankment of the River Reno. The Gallo weir is located approximately in the same place as the 1949 and 1951 breaches (Figure 11.1). This hydraulic structure allows controlled flooding of a certain area, where the expected economic damage is limited. The presence of the weir alters the safety level of the flood-prone area along the River Reno, which would be uniformly distributed if traditional resistance policies (see above) were used. In particular, the purpose of the weir is to increase the safety levels in the downstream flood-prone areas. Obviously, at the same time, the presence of the weir decreases the safety levels in the area affected by the controlled flooding, which notably is not a traditional retention basin. Hence, the Gallo weir can be considered a structural measure for implementing controlled flooding in certain areas (e.g. European Parliament, 2007). This flood risk management technique consists of allowing controlled inundation in areas where the impact of flood can then be minimized by ad-hoc non-structural measures, such as land-use management policies, in order to minimize the consequences of flooding.

This chapter describes a methodology, based on an ensemble of numerical simulations, to produce probability-weighted hazard maps (Di Baldassarre *et al.*, 2009c). The maps are used to investigate the effects of the weir on the distribution of inundation hazard within the entire study area.

11.3 METHODS

11.3.1 Hydraulic modelling

As mentioned in Chapter 5, the simulation of several inundation scenarios requires a compromise between physical realism and computational efficiency of the model. In this application example, numerical simulations were performed by using a hybrid methodology (Chapter 5): flows through the lateral weir and simulated breaches were computed by a 1D model and then adopted as the inflow boundary condition for a 2D model of the flood-prone area. In this approach, dynamic flooding can be simulated whilst avoiding the onerous description of the riverbed geometry in two dimensions, and consequently achieving a reduction in the computational time (Aureli *et al.*, 2006). In particular, this study used the 1D code HEC-RAS (Hydrologic Engineering Center, 2001) for simulating the hydraulic behaviour of the 28-km reach of the River Reno in the presence of levee breaches and with or without the Gallo weir, and the 2D code TELEMAC-2D (Galland *et al.*, 1991) for the floodplain flow.

Given that the Reno levees are 8–11 m high (see above), it can be reasonably assumed that there are no interactions between the two models (Aureli *et al.*, 2006). Thus, the two models were run separately: the output of the 1D model was used as input of the 2D model. The validity of this assumption was then verified for each run by analysing the evolution in time of the water levels simulated by the 2D model in the proximity of the breach.

The geometry of the 28-km reach of the River Reno was described by means of 35 cross sections, which was found to satisfy the guidelines for optimal cross-section spacing (see Chapter 4). The 1-in-100 year flood (Autorità di Bacino del Reno, 1998) was used as the upstream boundary condition, while a rating curve was used as the downstream boundary condition. The 1D model was calibrated by using hydrometric data (for a total of 120 points) referred to recent flood events (September 1994 and November 2000; see Autorità di Bacino del Reno, 2002).

The 2D model was used to simulate floodplain flow. After an extensive sensitivity analysis for identifying the optimal resolution (Chapter 5), the computational mesh was characterized by 9,437 elements and 4,885 nodes (cell size between 10 and 500 m). The altimetry was determined by means of the 10-m resolution DTM (Figure 11.1) and a survey of the principal breaklines, such as road embankments. The Manning's coefficients of the flood-prone area were selected to represent the physical characteristics of the flood-prone area according to standard tables (e.g. Chow, 1959) and sensitivity analysis of the 2D model previously performed in similar test sites (e.g. Horritt *et al.*, 2007). Then, in order to have an indication on the reliability of the model, the 2D model was tested by simulating the 1949 flood event. The agreement between the historical flood extent map (Figure 11.1) and the model simulation was around 85%.

11.3.2 Ensemble simulations

Flood hazard maps were generated by simulating several inundation scenarios, representing the absence of the Gallo weir (hypothetical scenario), or its presence (scenario representative of the current situation). More specifically, ensemble simulations were used to deal with the approximation induced by positioning the breach in different locations on the left levee and assuming different hypotheses of breach development. In this way it is possible to evaluate the uncertainty induced by the above unknown information. It is worth noting that the inundation modelling exercise is affected by other sources of uncertainty, such as imprecise input data (e.g. hydrologic input, river and floodplain geometry), model structural uncertainty and model parameters (see Chapter 6). In this application example, these other sources of uncertainty were neglected by assuming that these affect the inundation scenarios in a similar fashion and, therefore, that the results of the comparison are still suitable in the more general case.

11.4 RESULTS

11.4.1 Resistance strategy

As mentioned above, the hypothetical scenario neglects the presence of the Gallo weir, using a regular levee system instead (resistance strategy). This reflects the traditional resistance approach, where all flood-prone areas theoretically have the same probability of flooding.

To evaluate where levees might potentially be overtopped, a 1D preliminary simulation was carried out. In particular, the 1D model was used for simulating a synthetic flood event corresponding to a return period of 100 years (Autorità di Bacino del Reno, 1998). During this preliminary 1D simulation, flooding is restricted to the area inside the embankments (i.e. no overflow is allowed). The preliminary 1D simulation shows that the left levee system is not able to contain the 100-year flood (Figure 11.2).

This implies that the left levee system can be overtopped anywhere along the considered reach (Figure 11.2). Given that one cannot establish the exact location where the overtopping will start, 30 equally spaced potential locations for breaches along the River Reno were identified (Figure 11.3). Also, according to the description of historical levee breaches (Govi and Turitto, 2000),

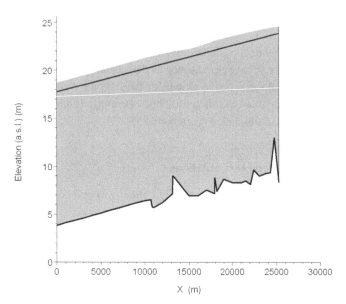

Figure 11.2 Absence of Gallo weir (resistance strategy): results of the preliminary 1D simulation in terms of maximum water depth (grey) and left levee elevation (upper black line).

the presence of only one breach along the levee system under study was assumed.

To account for the uncertainty associated with the location and evolution of the breach, an ensemble of numerical simulations was performed. In particular, for each breach location, this study considered different characteristics of the levee overtopping and breaching in terms of time of formation (overall duration of the breach development, T), width of the breach (W), and depth of the breach (D). The breach was assumed to start when the levee was overtopped. Given that it is impossible to determine a priori the value of T, W and D, a random generation of 200 different combinations of T, W and D was carried out under the assumption of uniform distribution of the three parameters and 1 hour $< T < 3$ hours, 100 m $< W < 300$ m, 0.5 m $< D < 4$ m (floodplain plan), according to historical data available for neighbouring sites (River Po; Govi and Turitto, 2000). Then, for each breach location, 200 1D simulations (each characterized by different T, W and D) were carried out by using the 100-year hydrograph as the upstream boundary condition. In total 200 breach outflow hydrographs were generated (Figure 11.4).

A distribution of hydrographs enables one to assess the uncertainty associated with the location and the evolution of the breach. In order to limit the number of 2D simulations, these outflows were statistically summarized by defining reference hydrographs for each breach location: high, medium and low, corresponding to the 75th percentile, the 50th percentile and the 25th percentile, respectively (Figure 11.5). A sensitivity analysis showed that the three reference hydrographs do not change significantly with a number of simulations larger than 100–150.

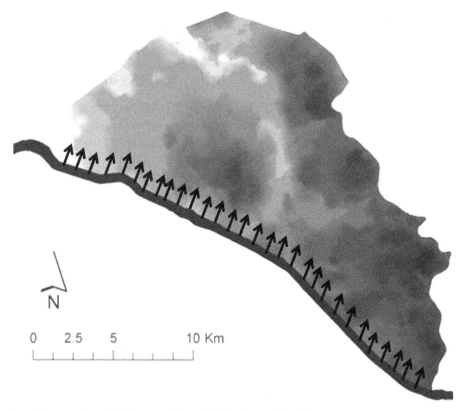

Figure 11.3 Location of possible levee breach (black arrows) along the River Reno (black line).

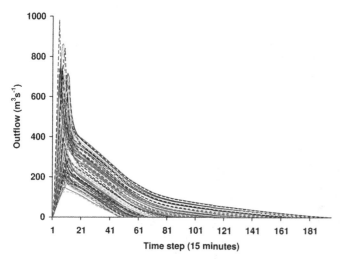

Figure 11.4 Levee breach: outflows obtained by means of 1D simulations for a particular location of the breach (Di Baldassarre *et al.*, 2009c).

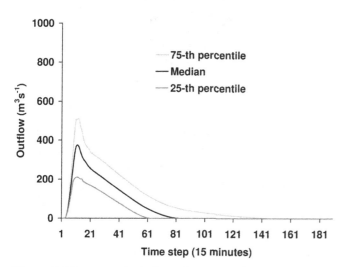

Figure 11.5 Representative hydrographs used as inflows to the 2D model (Di Baldassarre *et al.*, 2009c).

These reference hydrographs (Figure 11.5) were then used as the inflow condition for the 2D model to simulate inundation scenarios. The study used the 25th and 75th percentiles (rather than, for example, the 5th and 95th percentiles) in order to obtain plausible reference hydrographs and omit unlikely (i.e. extreme) combinations of the variables considered, which may originate as a result of the uniform distribution that is used to represent the frequency of the variables. If, for example, the 5th and 95th percentiles were used, unrepresentative hydrographs would be obtained. In total, in the case of the absence of the Gallo weir (resistance strategy), 6,000 1D unsteady flow simulations (30 different breach locations multiplied by 200 different breach characteristics) and 90 2D unsteady flow simulations for the overland flow (30 different breach locations multiplied by 3 representative inflow hydrographs) were performed.

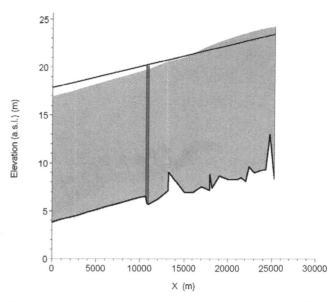

Figure 11.6 Presence of the Gallo weir: results of the preliminary 1D simulation in terms of maximum water depth (grey) and left levee elevation (upper black line).

11.4.2 Controlled flooding

The current situation scenario reflects the actual geometry with the presence of the Gallo weir, which allows controlled flooding of a certain prone area (controlled flooding strategy).

If the Gallo weir is included within the model (as a lateral structure; Hydrologic Engineering Center, 2001), the preliminary 1D simulation shows that, because of the controlled flooding through the lateral weir, the levee system is able to contain the 100-year flood downstream of the Gallo weir and for a short portion of the river upstream of the Gallo weir (Figure 11.6).

Thus, the left levee system can be overtopped at any point along the upstream part of the 28-km reach of the River Reno here considered, for a reach of around 9 km length (Figure 11.6). As mentioned above, one cannot establish the exact location where the overtopping will start. Therefore, 10 equally spaced potential locations for breaches along the upstream part of the 28-km reach of the River Reno were identified (Figure 11.7). Here, because of the presence of the Gallo weir, the inundation scenarios are characterized by a first inflow due to levee overtopping and breach and a second inflow due to the controlled flooding through the Gallo weir (Figure 11.7).

This study followed a procedure analogous to the procedure adopted for the resistance strategy in the absence of the Gallo weir (Section 11.4.1). In brief, in the presence of the Gallo weir, 2,000 1D unsteady flow simulations (10 different breach locations times 200 different breaches) and 30 2D unsteady flow simulations for the overland flow (10 different breach locations multiplied by 3 representative inflow hydrographs) were performed.

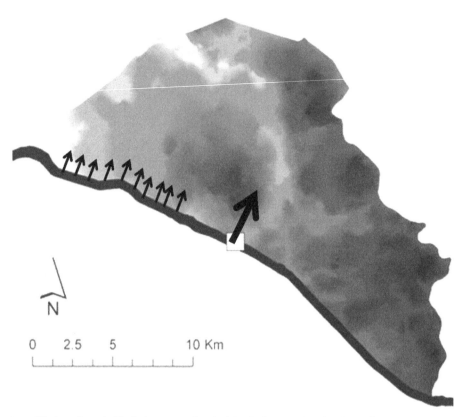

Figure 11.7 Location of possible levee breach (black short arrows) and of the Gallo controlled flooding (big arrow).

11.4.3 Flood hazard mapping

Flood hazard maps were generated by combining the results of the 2D simulations, in the absence or presence of the Gallo weir, and evaluating the expected water depth (WD_i) and the expected scalar velocity (SV_i), for each computational node i of the test site, as follows:

$$WD_i = \sum_{j=1}^{NS} x_j WD_{i,j} \qquad (11.1)$$

$$SV_i = \sum_{j=1}^{NS} x_j SV_{i,j} \qquad (11.2)$$

where NS is the number of 2D simulations. $WD_{i,j}$ and $SV_{i,j}$ are the maximum water depth and the maximum scalar velocity at the node i for the simulation j, and x_j is the weight of the simulation j obtained as a function of the probability of occurrence of the simulation itself, depending on the location and the magnitude of the breach:

$$x_j = wl_j wp_j \qquad (11.3)$$

where wl_j depends on the location of the breach and is linearly proportional to the difference between the maximum simulated water elevation and the levee system elevation (see Figures 11.2 and 11.6): the greater this difference, the higher the probability that the breach occurs (i.e. the levee system is more likely not to be able to contain the flood); wp_j depends on the magnitude of the breach and it was assumed to be equal to 0.25 for low

(L) or high (H) flows through the levee breach and it is equal to 0.50 for medium (M) flow (see below). This choice was made to give the same weight to medium and high or low flows. As a result, expected values of water depth and scalar velocity are obtained in view of the uncertainties due to the unknown position and development of the levee breach.

The scientific literature proposes several water depth–velocity hazard curves (e.g. ACER Technical Memorandum No. 11, 1988; Staatscourant, 1998; Vrisou van Eck and Kok, 2001). Despite these efforts, quantifying the expected flood damage is very difficult: the impact of a flood event on a prone area is related to several other factors such as the education of the population, the time of the day, and the day of the week when the inundation occurs; also, the consequences of a flood can last several months. For these reasons a flood hazard map was generated to evaluate the effects of the presence of the Gallo weir on the safety level of the flood-prone area. The study aimed at evaluating how the spatial distribution of hazards in the test site is affected by the presence of the Gallo weir; therefore it does not take into account other possible factors (e.g. land use). Figure 11.8 shows a possible relationship between flood depth, velocity and hazard. The curve represented in Figure 11.8 reflects those in the ACER Technical Memorandum No. 11 (1988). This relationship is appropriate for the test site under study as it was derived for permanent residences, commercial and public buildings, and worksite areas (ACER Technical Memorandum No. 11, 1988; Di Baldassarre *et al.*, 2009c).

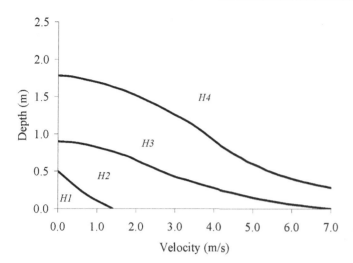

Figure 11.8 Flood hazard classification based on water depth and velocity.

Rigorously, the hazard level is to be evaluated by combining flood depth and velocity for each simulation at each time step. In order to make the procedure faster the value of the expected water depth and the expected scalar velocity were used for classifying the flood-prone area into five hazard classes: from H0, low flood hazard (corresponding to the area not concerned by inundation scenarios), to H4, very high flood hazard (Figure 11.8). In this way,

a first flood hazard map (Figure 11.9) was generated by combining the results of inundation scenarios corresponding to the hypothesis of regular levee system and a second map (Figure 11.10), by combining the results of the inundation scenarios corresponding to the implementation of controlled flooding, i.e. presence of the Gallo weir.

11.5 DISCUSSION

The modelling exercise presented here enabled the evaluation of the effects of controlled flooding strategies on the spatial distribution of inundation hazard. It is worth noting that the two inundation hazard maps (Figures 11.9 and 11.10) obtained by applying the simplified procedure are conservative. In fact, the proposed methodology tends to overestimate the hazard level as the timing of the maximum water depth is generally different from the timing of the maximum scalar velocity (Di Baldassarre *et al.*, 2009c).

By analysing the two maps one can observe that under the hypothesis of a regular levee system (i.e. resistance strategy), the inundation hazard is distributed over a wide area; whereas with the Gallo weir (i.e. controlled flooding), the hazard is more localized (Figures 11.9 and 11.10; see also Di Baldassarre *et al.*, 2009c). More specifically, with the presence of the Gallo weir,

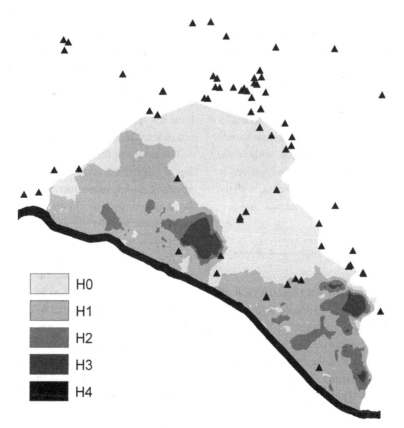

Figure 11.9 Flood hazard map (grey scale) under the hypothesis of absence of the Gallo weir and location of industrial plants (triangles) (Di Baldassarre *et al.*, 2009c).

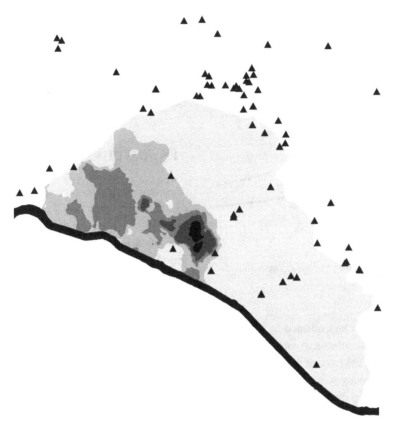

Figure 11.10 Flood hazard map (grey scale) under the hypothesis of presence of the Gallo weir (i.e. controlled flooding) and location of industrial plants (triangles) (Di Baldassarre *et al.*, 2009c).

flood hazard is increased in an area close to the Gallo weir, where hazard was already high with a regular levee system, because of the low terrain elevation (see DTM, Figure 11.1). Thus, the numerical exercise allowed the assessment of the effects of the hydraulic structure considered herein: due to controlled flooding, hazard slightly increases where it was already high and decreases significantly in a large part of the test site. This change should be regarded as an improvement since, for example, without the Gallo weir, many industrial plants and storage (see locations in Figure 11.9) would be located in H1 or H2 hazard zones; with the hydraulic structure they are all located in an H0 hazard zone.

11.6 CONCLUSIONS

This application example aimed to show the potential of flood hazard mapping, based on an ensemble of hydraulic simulations, to evaluate different flood mitigation techniques on a flood-prone area. By referring to an Italian case study, a numerical exercise was performed by generating two inundation hazard maps: one that neglects the presence of a lateral weir, using a regular levee system instead, and one that reflects the actual geometry with the presence

of a lateral weir, which allows controlled flooding. In particular, 1D and 2D flood inundation models were utilized and an innovative methodology for producing probability-weighted hazard maps based on ensembles of numerical simulations was described. The use of ensemble simulations allowed one to account for the most relevant sources of uncertainty in assessing inundation hazard in areas protected by river embankments.

The results of this exercise demonstrated that controlled flooding of large flood-prone areas – where damage is minimized by adopting ad-hoc non-structural measures – may be effective for flood risk mitigation. In particular, the study pointed out that hazard increases slightly where it was already high with a regular levee system, while it decreases significantly in a large part of the test site; this change should be regarded as an improvement, especially when considering the location of industrial plants and storage.

As a concluding remark, it is worth noting that the results of this application example are unavoidably associated with the considered test site. However, the methodology for producing probability-weighted hazard maps based on ensembles of numerical simulations can be used to compare alternative policies for flood risk management.

References

Abbott, M. B. (1979). *Computational Hydraulics: Elements of the Theory of Free Surface Flows*. London: Pitman.

Abbott, M. B., and Basco, D. R. (1989). *Computational Fluid Dynamics: An Introduction for Engineers*. Harlow, UK: Longman Scientific & Technical.

Abbott, M. B., and Ionescu, F. (1967). On the numerical computation of nearly horizontal flows. *Journal of Hydraulic Research*, **5**, 97–117.

ACER Technical Memorandum No. 11 (1988). *Downstream Hazard Classification Guidelines*. Assistant Commissioner – Engineering and Research, Denver, Colorado, US Department of the Interior, Bureau of Reclamation.

Ackerman, C. (2002). *HEC-GeoRAS: An Extension for Support of HEC-RAS Using ArcView GIS*. US Army Corps of Engineers.

Alsdorf, D. E., Smith, L. C., and Melack, J. M. (2001). Amazon floodplain water level changes measured with interferometric SIR-C radar. *IEEE Transactions on Geoscience and Remote Sensing*, **39**(2), 423–431.

Alsdorf, D. E., Bates, P. D., Melack, J., Wilson, M. D., and Dunne, T. (2007). The spatial and temporal complexity of the Amazon flood measured from space. *Geophysical Research Letters*, **34**, L08402.

Ambrosi, D. (1995). Approximation of shallow water equations by Roe's Riemann solver. *Journal for Numerical Methods in Fluids*, **20**, 157–168.

Apel, H., Aronica, G. T., Kreibich, H., and Thieken, A. H. (2009). Flood risk analyses: how detailed do we need to be? *Natural Hazards*, **49**, 79–98.

Aplin, P., Atkinson, P. M., Tatnall, A. R., Cutler, M. E., and Sargent, I. (1999). SAR imagery for flood monitoring and assessment. *Proceedings of the Remote Sensing Society, Earth Observation from Data to Information*, Cardiff, UK, 557–563.

Aronica, G., Hankin, B. G., and Beven, K. J. (1998). Uncertainty and equifinality in calibrating distributed roughness coefficients in a flood propagation model with limited data. *Advances in Water Resources*, **22**(4), 349–365.

Aronica, G., Bates, P. D., and Horritt, M. S. (2002). Assessing the uncertainty in distributed model predictions using observed binary pattern information within GLUE. *Hydrological Processes*, **16**(10), 2001–2016.

ARPA (2008). *Annali Idrologici: Parte Seconda* (in Italian). Agenzia Regionale Prevenzione e Ambiente, Regione Emilia Romagna, Servizio Idrometeorologico.

Ashworth, P. J., Bennett, S. J., Best, J. L., and McLelland, S. J. (1996). *Coherent Flow Structures in Open Channels*. Chichester, UK: John Wiley and Sons.

Audusse, E. (2005). A multilayer Saint-Venant model. *Discrete and Continuous Dynamical Systems, Series B*, **5**(2), 189–214.

Audusse, E., and Bristeau, M. O. (2007). Finite volume solvers for a multilayer Saint Venant system. *Journal of Applied Mathematical Computation*, **17**(3), 311–320.

Aureli, F., Mignosa, P., Ziveri, C., and Maranzoni, A. (2006). Fully-2D and quasi-2D modelling of flooding scenarios due to embankment failure. In *River Flow 2006*, London: Taylor & Francis Group.

Aureli, F., Maranzoni, A., Mignosa, P., and Ziveri, C. (2008). Dam-break flows: acquisition of experimental data through an imaging technique and 2D numerical modeling. *Journal of Hydraulic Engineering*, **134**, 1089.

Autorità di Bacino del Reno (1998). *Generazione di idrogrammi di piena nel bacino del fiume Reno chiuso a Casalecchio* (in Italian). Published online at www.regione.emilia-romagna.it/bacinoreno, Bologna.

Autorità di Bacino del Reno (2002). *Piano stralcio per l'assetto idrogeologico* (in Italian). Published online at www.regione.emilia-romagna.it/bacinoreno, Bologna.

Barkau, R. L. (1997). *UNET: One-dimensional Unsteady Flow Through a Full Network of Open Channels. User's Manual*. Davis, CA: US Army Corps of Engineering, Hydrologic Engineering Center.

Bates, B. C., Kundzewicz, Z. W., Wu, S., and Palutikof, J. P. (2008). *Climate Change and Water*. Geneva: Intergovernmental Panel on Climate Change Secretariat, Technical Paper.

Bates, P. D. (2004a). Remote sensing and flood inundation modelling. *Hydrological Processes*, **18**, 2593–2597.

Bates, P. D. (2004b). Computationally efficient modelling of flood inundation extent. In Brath, A., Montanari, A., and Toth, E. (eds.), *Hydrological Risk*, Cosenza, Italy: BIOS, 285–301.

Bates, P. D., and De Roo, A. P. J. (2000). A simple raster based model for flood inundation simulation. *Journal of Hydrology*, **236**, 54–77.

Bates, P. D., and Horritt, M. S. (2005). Modelling wetting and drying processes in hydraulic models. In Bates, P. D., Lane, S. N., and Ferguson, R. I. (eds.), *Computational Fluid Dynamics: Applications in Environmental Hydraulics*, Chichester, UK: John Wiley and Sons.

Bates, P. D., Stewart, M. D., Siggers, G. B., *et al.* (1998). Internal and external validation of a two-dimensional finite element model for river flood simulation. *Proceedings of the Institution of Civil Engineers, Water Maritime and Energy*, **130**, 127–141.

Bates, P. D., Stewart, M. D., Desitter, A., *et al.* (2000). Numerical simulation of floodplain hydrology. *Water Resources Research*, **36**, 2517–2530.

Bates, P. D., Marks, K. J., and Horritt, M. S. (2003). Optimal use of high-resolution topographic data in flood inundation models. *Hydrological Processes*, **17**(3), 537–557.

Bates, P. D., Horritt, M. S., Aronica, G., and Beven, K. (2004). Bayesian updating of flood inundation likelihoods conditioned on flood extent data. *Hydrological Processes*, **18**, 3347–3370.

Bates, P. D., Horritt, M. S., Hunter, N. M., Mason, D., and Cobby, D. (2005). Numerical modelling of floodplain flow. In Bates, P. D., Lane, S. N., and Ferguson, R. I. (eds.), *Computational Fluid Dynamics: Applications in Environmental Hydraulics*, Chichester: John Wiley and Sons, 271–304.

Bates, P. D., Wilson, M. D., Horritt, M. S., *et al.* (2006). Reach scale floodplain inundation dynamics observed using airborne Synthetic Aperture Radar imagery: data analysis and modelling. *Journal of Hydrology*, **328**, 306–318.

Bates, P. D., Horritt, M. S., and Fewtrell, T. J. (2010). A simple inertial formulation of the shallow water equations for efficient two-dimensional flood inundation modelling. *Journal of Hydrology*, **387**, 33–45.

Begnudelli, L., Sanders, B. F., and Bradford, S. F. (2008). Adaptive Godunov-based model for flood simulation. *Journal of Hydraulic Engineering*, **134**(6), 714–725.

Beven, K. J. (1989). Changing ideas in hydrology: the case of physically-based models. *Journal of Hydrology*, **105**, 157–172.

Beven, K. J. (1995). Linking parameters across scales: subgrid parameterizations and scale-dependent hydrological models. *Hydrological Processes*, **9**(5–6), 507–525.

Beven, K. J. (2000). Uniqueness of place and process representations in hydrological modeling. *Hydrology and Earth System Sciences*, **4**(2), 203–213.

Beven, K. J. (2001). *Rainfall Runoff Modelling: The Primer*. Chichester, UK: John Wiley and Sons.

Beven, K. J. (2002). Towards a coherent philosophy for modelling the environment. *Proceedings of the Royal Society of London Series A: Mathematical Physical and Engineering Sciences*, **458** (2026), 2465–2484.

Beven, K. J. (2006). A manifesto for the equifinality thesis. *Journal of Hydrology*, **320**(1–2), 18–36.

Beven, K. J. (2008). On doing better hydrological science. *Hydrological Processes*, **22**, 3549–3553.

Beven, K. J., and Binley, A. M. (1992). The future of distributed models: model calibration and uncertainty prediction. *Hydrological Processes*, **6**, 279–298.

Beven, K. J., and Freer, J. (2001). Equifinality, data assimilation, and uncertainty estimation in mechanistic modelling of complex environmental systems. *Journal of Hydrology*, **249**, 11–29.

Bloeschl, G. (2006). Hydrologic synthesis: across processes, places, and scales. *Water Resources Research*, **42**, W03S02.

Blöschl, G., and Montanari, A. (2010). Climate change impacts –throwing the dice? *Hydrological Processes*, **24**, 374–381.

Blyth, K. (1997). FLOODNET: a telenetwork for acquisition, processing, and dissemination of Earth Observation data for monitoring and emergency management of floods. *Hydrological Processes*, **11**, 1359–1375.

Box, G. E. P., and Jenkins, G. M. (1970). *Time Series Analysis: Forecasting and Control*. San Francisco, USA: Holden Day Press.

Brakenridge, G. R., Tracy, B. T., and Knox, J. C. (1998). Orbital SAR remote sensing of a river flood wave. *International Journal of Remote Sensing*, **19**(7), 1439–1445.

Brandimarte, L., Brath, A., Castellarin, A., and Di Baldassarre, G. (2009). Isla Hispaniola: a trans-boundary flood risk mitigation plan. *Physics and Chemistry of the Earth*, **34**, 209–218.

Braschi, G., Gallati, M., and Natale, L. (1989). Simulation of a road network flooding. *20th Annual Pittsburgh Conference on Modeling and Simulation*, Pittsburgh, USA, **4**, 1625–1632.

Brath, A., and Di Baldassarre, G. (2006). Modelli matematici per l'analisi della sicurezza idraulica del territorio (in Italian). *L'Acqua*, **6**, 39–48.

Bridge, J. S., and Gabel, S. L. (1992). Flow and sediment dynamics in a low sinuosity, braided river: Calamus River, Nebraska sandhills. *Sedimentology*, **39**(1), 125–142.

BRISK (2011). Bristol Environmental Risk Research Centre (www.bristol.ac.uk/brisk/research).

Brookes, A. N., and Hughes, T. J. R. (1982). Streamline Upwind/Petrov Galerkin formulations for convection dominated flows with particular emphasis on the incompressible Navier–Stokes equations. *Computer Methods in Applied Mechanics and Engineering*, **32**, 199–259.

Burnham, K. P., and Anderson, D. R. (2002). *Model Selection and Multimodel Inference*, 2nd edition. New York: Springer.

Burnham, M. W., and Davis, D. W. (1990). Effects of data errors in computed steady-flow profiles. *Journal of Hydraulic Engineering*, **116**, 914–928.

Burrough, P. A. (1998). Dynamic modelling and geocomputation. In Longley, P. A., Brooks, S. M., McDonnell, R. M., and Macmillan, B. (eds.), *Geocomputation*, Chichester: Wiley, 165–191.

Burton, C., and Cutter, S. L. (2008). Levee failures and social vulnerability in the Sacramento–San Joaquin Delta area, California. *Natural Hazards Review*, **9**(3), 136–149.

Cameron, D. S., Beven, K. J., Tawn, J., Blazkova, S., and Naden, P. (1999). Flood frequency estimation by continuous simulation for a gauged upland catchment (with uncertainty). *Journal of Hydrology*, **219**(3–4), 169–187.

Camorani, G., Castellarin, A., and Brath, A. (2006). Effects of land-use changes on the hydrologic response of reclamation systems. *Physics and Chemistry of the Earth*, **30**, 561–574.

Castellarin, A., Di Baldassarre, G., Bates, P. D., and Brath, A. (2009). Optimal cross-section spacing in Preissmann scheme 1D hydrodynamic models. *ASCE Journal of Hydraulic Engineering*, **135**(2), 96–105.

Castellarin, A., Di Baldassarre, G., and Brath, A. (2011). Floodplain management strategies for flood attenuation in the River Po. *River Research and Applications*, **27**, 1037–1047.

Casulli, V. (1990). Semi-implicit finite difference methods for the two-dimensional shallow water equations. *Journal of Computational Physics*, **86**, 56–74.

Casulli, V., and Zanolli, P. (2002). Semi-implicit numerical modeling of non-hydrostatic free-surface flows for environmental problems. *Mathematical and Computer Modelling*, **36**,1131–1149.

Chow, V. T. (1959). *Open Channel Hydraulics*. New York: McGraw-Hill.

Chow, V. T., Maidment, D. R., and Mays, L.W. (1998). *Applied Hydrology*. New York: McGraw-Hill.

Church, J. A., and White, N. J. (2011). Sea-level rise from the late 19th to the early 21st century. *Surveys in Geophysics*, **32**, 585–602.

Clarke, D. (2005). *Managing Flood Risk: Dealing with Flooding*. Technical Report GEHO0605BJDB-E-E, Environment Agency.

Clarke, R. T. (1999). Uncertainty in the estimation of mean annual flood due to rating-curve indefinition. *Journal of Hydrology*, **222**, 185–190.

Cloke, H. L., and Hannah, D. M. (2011). Large-scale hydrology: advances in understanding processes, dynamics and models from beyond river basin to global scale. *Hydrological Processes*, **25**, 991–995.

Cobby, D. M., Mason, D. C., and Davenport, I. J. (2001). Image processing of airborne scanning laser altimetry data for improved river flood modeling. *ISPRS Journal of Photogrammetry and Remote Sensing*, **56**(2), 121–138.

Cobby, D. M., Mason, D. C., Horritt, M. S., and Bates, P. D. (2003). Two-dimensional hydraulic flood modelling using a finite-element mesh decomposed according to vegetation and topographic features derived from airborne scanning laser altimetry. *Hydrological Processes*, **17**(10), 1979–2000.

Coratza, L. (2005). *Aggiornamento del Catasto delle Arginature Maestre di Po* (in Italian). Parma, Italy: Po River Basin Authority.

Costanza, R., d'Arge, R., de Groot, R., *et al.* (1997). The value of the world's ecosystem services and natural capital. *Nature*, **387**, 253–260.

Cunge, J. A. (1969). On the subject of flood propagation computation method (Muskingum method). *Journal of Hydraulic Research*, **7**(2), 205–230.

Cunge, J. A. (2003). Of data and models. *Journal of Hydroinformatics*, **5**, 75–98.

Cunge, J. A., Holly, F. M., and Verwey, A. (1980). *Practical Aspects of Computational River Hydraulics*. London: Pitman.

Dartmouth Flood Observatory (2010). *Global Archive of Large Flood Events*. Available at www.dartmouth.edu/~floods.

Day, A.-L. (2005). *Carlisle Storms and Associated Flooding: Multi-Agency Debrief Report*. Technical Report, UK Resilience.

De Bruijn, K. M., and Klijn, F. (2001). Resilient flood risk management strategies. In Guifen, L., and Wenxue, L. (eds.), *Proceedings of the IAHR Congress*, Beijing, China: Tsinghua University Press, 450–457.

Defina, A. (2000). Two-dimensional shallow flow equations for partially dry areas. *Water Resources Research*, **36**(11), 3251–3264.

Di Baldassarre, G., and Claps, P. (2011). A hydraulic study on the applicability of flood rating curves. *Hydrology Research*, **42**(1), 10–19.

Di Baldassarre, G., and Montanari, A. (2009). Uncertainty in river discharge observations: a quantitative analysis. *Hydrology and Earth System Sciences*, **13**, 913–921.

Di Baldassarre, G., and Uhlenbrook, S. (2011). Is the current flood of data enough? A treatise on research needs to improve flood modelling. *Hydrological Processes*, doi: 10.1002/hyp.8226.

Di Baldassarre, G., Schumann, G., and Bates, P. D. (2009a). Near real time satellite imagery to support and verify timely flood modelling. *Hydrological Processes*, **23**, 799–803.

Di Baldassarre, G., Schumann, G., and Bates, P. D. (2009b). A technique for the calibration of hydraulic models using uncertain satellite observations of flood extent. *Journal of Hydrology*, **367**, 276–282.

Di Baldassarre, G., Castellarin, A., Montanari, A., and Brath, A. (2009c). Probability weighted hazard maps for comparing different flood risk management strategies: a case study. *Natural Hazards*, **50**(3), 479–496.

Di Baldassarre, G., Laio, F., and Montanari, A. (2009d). Design flood estimation using model selection criteria. *Physics and Chemistry of the Earth*, **34**(10–12), 606–611.

Di Baldassarre, G., Castellarin, A., and Brath, A. (2009e). Analysis on the effects of levee heightening on flood propagation: some thoughts on the River Po. *Hydrological Sciences Journal*, **54**(6), 1007–1017.

Di Baldassarre, G., Schumann, G., Bates, P., Freer, J., and Beven, K. (2010a). Floodplain mapping: a critical discussion on deterministic and probabilistic approaches. *Hydrological Sciences Journal*, **55**(3), 364–376.

Di Baldassarre, G., Montanari, A., Lins, H., *et al.* (2010b). Flood fatalities in Africa: from diagnosis to mitigation. *Geophysical Research Letters*, **37**, L22402, doi:10.1029/2010GL045467.

Di Baldassarre, G., Schumann, G., Brandimarte, L., and Bates, P. D. (2011a). Timely low resolution SAR imagery to support floodplain modelling: a case study review. *Surveys in Geophysics*, **32**(3), 255–269.

Di Baldassarre, G., Elshamy, M., van Griensven, A., *et al.* (2011b). Future hydrology and climate in the River Nile basin: a review. *Hydrological Sciences Journal*, **56**(2), 199–211.

Dottori, F., and Todini, E. (2011). Developments of a flood inundation model based on the cellular automata approach: testing different methods to improve model performance. *Physics and Chemistry of the Earth*, **36**, 266–280.

Dottori, F., Martina, M. L. V., and Todini, E. (2009). A dynamic rating curve approach to indirect discharge measurement. *Hydrology and Earth System Sciences*, 6, 859–896.

Eilertsen, R. S., and Hansen, L. (2008). Morphology of river bed scours on a delta plain revealed by interferometric sonar. *Geomorphology*, 94(1–2), 58–68.

EM-DAT (Emergency Events Database) (2010). *OFDA/CRED International Disaster Database*, Universite Catholique de Louvain, Brussels, www.cred.be/emdat.

Environment Agency (2003). *River Dee Catchment Flood Management Plan, Hydrological and Hydraulic Modelling Report, Final Report*. Environment Agency Wales, Buckley, Flintshire, UK.

Ervine, D. A., and Baird, J. I. (1982). Rating curves for rivers with overbank flow. *Proceedings of the Institution of Civil Engineers Part 2: Research and Theory*, 73, 465–472.

European ISO EN Rule 748 (1997). *Measurement of Liquid Flow in Open Channels: Velocity–Area Methods*. Reference number ISO 748:1997 (E), International Standard.

European Parliament (2007). *Directive 2007/60/EC of the European Parliament and the Council of October 2007 on the Assessment and Management of Flood Risks*. European Floods Directive, L 288/27, Official Journal of the European Union, Brussels, available at http://ec.europa.eu/environment/water/flood_risk/index.htm.

Fenicia, F., Savenije, H. H. G., Matgen, P., and Pfister, L. (2008). Understanding catchment behavior through stepwise model concept improvement. *Water Resources Research*, 44, W01402, 1–13.

Fewtrell, T. J., Bates, P. D., Horritt, M., and Hunter, N. M. (2008). Evaluating the effect of scale in flood inundation modelling in urban environments. *Hydrological Processes*, 22(26), 5107–5118.

Fisher, K., and Dawson, H. (2003). *Reducing Uncertainty in River Flood Conveyance: Roughness Review*. UK DEFRA and Environment Agency Report, W5A-057, Environment Agency, Bristol, UK.

Franchini, M., Lamberti, P., and Di Giammarco, P. (1999). Rating curve estimation using local stages, upstream discharge data and a simplified hydraulic model, *Hydrology and Earth System Sciences*, 3, 541–548.

Freer, J., and Beven, K. (2005). Model structural error and the curse of the errors in variables problem. *EGU General Assembly*, Vienna, Austria.

Freer, J., Beven, K. J., and Ambroise, B. (1996). Bayesian estimation of uncertainty in runoff prediction and the value of data: an application of the GLUE approach. *Water Resources Research*, 32, 2163–2173.

Galland, J. C., Goutal, N., and Hervouet, J. M. (1991). TELEMAC: A new numerical model for solving shallow water equations. *Advances in Water Resources*, 14(3), 38–148.

Galloni, E. (1881). *Cenni monografici sui singoli servizi dipendenti dal Ministero dei Lavori Pubblici per gli anni 1878–1879–1880 compilati in occasione dell'Esposizione Nazionale di Milano dell'anno 1881* (Catalogue of the activities of the Ministry of Public Works in the period 1878–1880, reported during the 1881 National Exposition in Milan, in Italian). Rome, Italy: Eredi Botta.

Gerbeau, J.-F., and Perthame, B. (2001). Derivation of viscous Saint-Venant system for laminar shallow water, numerical validation. *Discrete and Continuous Dynamical Systems, Series B*, 1(1), 89–102.

Götzinger, J., and Bardossy, A. (2008). Generic error model for calibration and uncertainty estimation of hydrological models. *Water Resources Research*, 44, W00B07, doi:10.1029/2007WR006691.

Govi, M., and Turitto, O. (2000). *Casistica storica sui processi d'iterazione delle correnti di piena del Po con arginature e con elementi morfotopografici del territorio adiacente* (Historical documentation about the processes of dam breaks in the River Po, in Italian). Istituto Lombardo Accademia di Scienza e Lettere.

Gupta, H. V., Beven, K. J., and Wagener, T. (2005). Model calibration and uncertainty estimation. In *Encyclopedia of Hydrological Sciences*, Anderson, M. G. (ed.), New York: John Wiley, 2015–2031.

Gupta, R. P., and Banerji, S. (1985). Monitoring of reservoir volume using LANDSAT data. *Journal of Hydrology*, 77, 159–170.

Hall, J., and Beven, K. (2011). *Applied Uncertainty Analysis for Flood Risk Management*. London: Imperial College Press.

Hall, J. W., Sayers, P. B., and Dawson, R. J. (2005a). National-scale assessment of current and future flood risk in England and Wales. *Natural Hazards*, 36, 147–164.

Hall, J. W., Tarantolo, S., Bates, P., and Horritt, M. S. (2005b). Distributed sensitivity analysis of flood inundation model calibration. *Journal of Hydraulic Engineering*, 131(2), 117–126.

Hankin, B. G., Hardy, R., Kettle, H., and Beven, K. J. (2001). Using CFD in a GLUE framework to model the flow and dispersion characteristics of a natural fluvial dead zone. *Earth Surface Processes and Landforms*, 26(6), 667–687.

Hardy, R. J., Bates, P. D., and Anderson, M. G. (1999). The importance of spatial resolution in hydraulic models for floodplain environments. *Journal of Hydrology*, 216(1–2), 124–136.

Hereher, M. E. (2010). Vulnerability of the Nile Delta to sea level rise: an assessment using remote sensing. *Geomatics, Natural Hazards and Risk*, 1(4), 315–321.

Herschy, R. W. (1978). *Accuracy in Hydrometry*. New York: Wiley.

Hervouet, J.-M., and Van Haren, L. (1996). Recent advances in numerical methods for fluid flows. In Anderson, M. G., Walling, D. E., and Bates, P. D. (eds.), *Floodplain Processes*, Chichester, UK: John Wiley and Sons, 183–214.

Holling, C. S. (1973). Resilience and stability of ecological systems. *Annual Review of Ecology and Systematics*, 4, 1–24.

Horritt, M. S. (2000). Development of physically based meshes for two-dimensional models of meandering channel flow. *International Journal for Numerical Methods in Engineering*, 47, 2019–2037.

Horritt, M. S. (2005). Parameterisation, validation and uncertainty analysis of CFD models of fluvial and flood hydraulics in the natural environment. In Bates, P. D., Lane, S. N., and Ferguson, R. I. (eds.), *Computational Fluid Dynamics: Applications in Environmental Hydraulics*. Chichester, UK: John Wiley and Sons, 193–214.

Horritt, M. S. (2006). A methodology for the validation of uncertain flood inundation models. *Journal of Hydrology*, 326, 153–165.

Horritt, M. S., and Bates, P. D. (2001). Effects of spatial resolution on a raster based model of flood flow. *Journal of Hydrology*, 253, 239–249.

Horritt, M. S., and Bates, P. D. (2002). Evaluation of 1-D and 2-D models for predicting river flood inundation. *Journal of Hydrology*, 268, 87–99.

Horritt, M. S., Mason, D., and Luckman, A. J. (2001). Flood boundary delineation from synthetic aperture radar imagery using a statistical active contour model. *International Journal of Remote Sensing*, 22, 2489–2507.

Horritt, M. S., Mason, D. C., Cobby, D. M., Davenport, I. J., and Bates, P. D. (2003). Waterline mapping in flooded vegetation from airborne SAR imagery. *Remote Sensing of Environment*, 85(3), 271–281.

Horritt, M. S., Bates, P. D., and Mattinson, M. J. (2006). Effects of mesh resolution and topographic representation in 2D finite volume models of shallow water fluvial flow. *Journal of Hydrology*, 329(1–2), 306–314.

Horritt, M. S., Di Baldassarre, G., Bates, P. D., and Brath, A. (2007). Comparing the performance of 2-D finite element and finite volume models of floodplain inundation using airborne SAR imagery. *Hydrological Processes*, 21, 2745–2759.

Horritt, M. S., Bates, P., Fewtrell, T., Mason, D., and Wilson, M. (2010). Modelling the hydraulics of the Carlisle 2005 flood event. *Proceedings of the Institution of Civil Engineers: Water Management*, 163, 273–281.

Hostache, R., Matgen, P., Schumann, G., *et al.* (2009). Water level estimation and reduction of hydraulic model calibration uncertainties using satellite SAR images of floods. *IEEE Transactions on Geoscience and Remote Sensing*, 47, 431–441.

Hughes, D., Greenwood, P., Coulson, G., and Blair, G. (2007). GridStix: supporting flood prediction using embedded hardware and next generation grid middleware. *Proceedings of the 2006 International Symposium on a World of Wireless, Mobile and Multimedia Networks*, IEEE Computer Society.

Hunter, N. M., Bates, P. D., Horritt, M. S., *et al.* (2005a). Utility of different data types for calibrating flood inundation models within a GLUE framework. *Hydrology and Earth System Sciences*, 9(4), 412–430.

Hunter, N. M., Horritt, M. S., Bates, P. D., Wilson, M. D., and Werner, M. G. F. (2005b). An adaptive time step solution for raster-based storage cell modelling of floodplain inundation. *Advances in Water Resources*, 28(9), 975–991.

Hunter, N. M., Bates, P. D., Horritt, M. S., *et al.* (2007). Simple spatially-distributed models for predicting flood inundation: a review. *Geomorphology*, 90, 208–225.

Hunter, N. M., Bates, P. D., Neelz, S., *et al.* (2008). Benchmarking 2D hydraulic models for urban flooding. *Proceedings of the Institution of Civil Engineers: Water Management*, 161, 13–30.

Hydrologic Engineering Center (2001). *Hydraulic Reference Manual*. Davis, CA: US Army Corps of Engineers.

Institute of Hydrology (1999). *Flood Estimation Handbook*. Wallingford, UK: Institute of Hydrology.

tersen, G. W. (1981). Texture transforms of remote sensing. _Sensing of Environment_, **11**, 359–370.

_, J. P. F. M., and Jorissen, R. E. (1997). Flood management in the Netherlands: recent developments and research needs. In Casale, R., Havno, K., and Samuels, P. (eds.), _Ribamod, River Basin Modelling, Management and Flood Mitigation, Concerted Action_, London: Taylor & Francis Group, 89–104.

Jarret, R. D. (1987). Errors in slope–area computations of peak discharges in mountain streams. _Journal of Hydrology_, **96**, 53–67.

Jones, B. E. (1916). A method of correcting river discharge for a changing stage. _US Geological Survey Water Supply Paper_, **375-E**, 117–130.

Kirby, W. H. (1987). Linear error analysis of slope–area discharge determinations. _Journal of Hydrology_, **96**, 125–138.

Knight, D. W., and Shiono, K. (1996). River channel and floodplain hydraulics. In Anderson, M. G., Walling, D. E., and Bates, P. D. (eds.), _Floodplain Processes_, Chichester, UK: John Wiley and Sons, 139–182.

Kohane, R., and Welz, R. (1994). Combined use of FE models for prevention of ecological deterioration of areas next to a river hydropower complex. In Peter, A., Wittum, G., Meissner, U., _et al._ (eds.), _Computational Methods in Water Resources_, Volume **1**, Dordrecht, the Netherlands: Kluwer, 59–66.

Komma, J., and Blöschl, G. (2008). Potential of non-structural flood mitigation measures. _Geophysical Research Abstracts_, EGU General Assembly 2008, **10**, EGU2008-A-08341.

Kussul, N., Shelestov, A., and Skakun, S. (2008). Grid system for flood extent extraction from satellite images. _Earth Science Informatics_, **1**(3), 105–117.

Laio, F. (2004). Cramer–von Mises and Anderson–Darling goodness of fit tests for extreme value distributions with unknown parameters. _Water Resources Research_, **40**, W09308, doi:10.1029/2004WR003204.

Laio, F., Di Baldassarre, G., and Montanari, A. (2009). Model selection techniques for the frequency analysis of hydrological extremes. _Water Resources Research_, **45**, W07416.

Lamb, R., Crossley, A., and Waller, S. (2009). A fast 2D floodplain inundation model. _Proceedings of the Institution of Civil Engineers: Water Management_, **162**(6), 363–370.

Landis, W. G. (ed.) (2005). _Regional Scale Ecological Risk Assessment Using the Relative Risk Model_. Boca Raton, FL, USA: CRC Press.

Lane, S. N. (2005). Roughness: time for a re-evaluation? _Earth Surface Processes and Landforms_, **30**(2), 251–253.

Lane, S. N., and Hardy, R. J. (2002). Porous rivers: a new way of conceptualising and modelling river and floodplain flows. In Ingham, D. B., and Pop, I. (eds.), _Transport Phenomena in Porous Media II_, Oxford, UK: Pergamon Press, 425–449.

Lane, S. N., James, T. D., Pritchard, H., and Saunders, M. (2003). Photogrammetric and laser altimetric reconstruction of water levels for extreme flood event analysis. _Photogrammetric Record_, **18**(104), 293–307.

LeFavour, G., and Alsdorf, D. (2005). Water slope and discharge in the Amazon river estimated using the shuttle radar topography mission digital elevation model. _Geophysical Research Letters_, **32**, L17404.

Léonard, J., Mietton, M., Najib, H., and Gourbesville, P. (2000). Rating curve modelling with Manning's equation to manage instability and improve extrapolation. _Hydrological Sciences Journal_, **45**(5), 739–750.

LeVeque, R. (2002). _Finite Volume Methods for Hyperbolic Problems_. Cambridge, UK: Cambridge University Press.

Liang, D., Lin, B., and Falconer, R. A. (2007). Simulation of rapidly varying flow using an efficient TVD–MacCormack scheme. _International Journal for Numerical Methods in Fluids_, **53**(5), 811–826.

Lighthill, M. G., and Whitham, G. B. (1955). On kinematic waves. II: A theory of traffic flow on long crowded roads. _Proceedings of the Royal Society_, **229**, 317–345.

Lincoln, T. (2007). Hydrology: flood of data. _Nature_, **447**, 393.

Lintrup, M. (1989). A new expression for the uncertainty of a current meter discharge measurement. _Nordic Hydrology_, **20**(3), 191–200.

Lynch, D. R., and Gray, W. G. (1980). Finite element simulation of flow deforming regions. _Journal of Computational Physics_, **36**, 135–153.

Maione, U., Mignosa, P., and Tomirotti, M. (2003). Regional estimation model of synthetic design hydrographs. _International Journal of River Basin Management_, **12**, 151–163.

Mantovan, P., and Todini, E. (2006). Hydrological forecasting uncertainty assessment: incoherence of the GLUE methodology. _Journal of Hydrology_, **330**, 368–381.

Marchuk, G. (1975). _Methods of Numerical Mathematics_. New York: Springer-Verlag.

Mark, O., Weesakul, S., Apirumanekul, C., Aroonnet, S. B., and Djordjevic, S. (2004). Potential and limitations of 1d modelling of urban flooding. _Journal of Hydrology_, **299**(3–4), 284–299.

Marks, K., and Bates, P. D. (2000). Integration of high-resolution topographic data with floodplain flow models. _Hydrological Processes_, **14**(11–12), 2109–2122.

Mason, D. C., Davenport, I. J., Flather, R. A., _et al._ (2001). A sensitivity analysis of the waterline method of constructing a digital elevation model for intertidal areas in ERS SAR scene of Eastern England. _Estuarine, Coastal and Shelf Science_, **53**, 759–778.

Mason, D. C., Cobby, D. M., Horritt, M. S., and Bates, P. D. (2003). Floodplain friction parameterization in two-dimensional river flood models using vegetation heights derived from airborne scanning laser altimetry. _Hydrological Processes_, **17**, 1711–1732.

Mason, D. C., Horritt, M. S., Bates, P. D., and Hunter, N. (2007). Use of fused airborne scanning laser altimetry and digital map data for urban flood modelling. _Hydrological Processes_, **21**(11), 1436–1447.

Mason, D. C., Bates, P. D., and Dall'Amico, J. T. (2009). Calibration of uncertain flood inundation models using remotely sensed water levels. _Journal of Hydrology_, **368**, 224–236.

Mason, D. C., Speck, R., Devereux, B., _et al._ (2010). Flood detection in urban areas using TerraSAR-X. _IEEE Transactions on Geoscience and Remote Sensing_, **48**(2), 882–894.

Matgen, P., Schumann, G., Henry, J. B., Hoffmann, L., and Pfister, L. (2007). Integration of SAR-derived inundation areas, high precision topographic data and a river flow model toward real-time flood management. _International Journal of Applied Earth Observation and Geoinformation_, **9**(3), 247–263.

McMillan, H. K., and Brasington, J. (2007). Porosity techniques and flooding in Cambridge. _Geomorphology_, **90**, 226–243.

Merwade, V., Olivera, F., Arabi, M., and Edleman, S. (2008). Uncertainty in flood inundation mapping: current issues and future directions. _Journal of Hydrologic Engineering_, **13**(7), 608–620.

Merz, B., Thieken, A. H., and Gocht, M. (2007). Flood risk mapping at the local scale: concepts and challenges. In Begum, S., Stive, M. J. F., and Hall, J. W. (eds.), _Flood Risk Management in Europe: Innovation in Policy and Practice_, Dordrecht, the Netherlands: Springer, 231–251.

Mignot, E., Paquier, A., and Haider, S. (2006). Modeling floods in a dense urban area using 2D shallow water equations. _Journal of Hydrology_, **327**(1–2), 186–199.

Mitkova, V., Pekarova, P., Miklanek, P., and Pekar, J. (2005). Analysis of flood propagation changes in the Kienstock–Bratislava reach of the Danube River. _Hydrological Sciences Journal_, **50**(4), 655–668.

Mitosek, H. T., Strupczewski, W. G., and Singh, V. P. (2006). Three procedures for selection of annual flood peak distribution. _Journal of Hydrology_, **323**, 57–73.

Montanari, A. (2005). Large sample behaviors of the generalized likelihood uncertainty estimation (GLUE) in assessing the uncertainty of rainfall-runoff simulations. _Water Resources Research_, **41**(8), WR08406.

Montanari, A. (2007). What do we mean by uncertainty? The need for a consistent wording about uncertainty assessment in hydrology. _Hydrological Processes_, **21**, 841–845, doi:10.1002/hyp.6623.

Mujumdar, P., and Kumar, D. N. (2012). _Floods in a Changing Climate: Hydrological Modeling_. International Hydrology Series, Cambridge, UK: Cambridge University Press.

Nardi, F., Vivoni, E. R., and Grimaldi, S. (2006). Investigating a floodplain scaling relation using a hydrogeomorphic delineation method. _Water Resources Research_, **42**, W09409.

Nash, J. E., and Sutcliffe, J. V. (1970). River flow forecasting through conceptual models. Part I: A discussion of principles. _Journal of Hydrology_, **10**(3), 282–290.

Natale, L., Savi, F., and Ubertini, L. (2002). Flood wave propagation: effect of the river geometry modification. In _Proceedings IASTED International Conference on Applied Simulation and Modelling_, 25–28 June 2002, Crete, Greece, 519–523.

Neal, J., Bates, P., Fewtrell, T., _et al._ (2009a). Hydrodynamic modelling of the Carlisle 2005 urban flood event and comparison with validation data. _Journal of Hydrology_, **375**, 589–600.

Neal, J., Fewtrell, T., and Trigg, M. (2009b). Parallelisation of storage cell flood models using OpenMP. _Environmental Modelling and Software_, **24**, 872–877.

Neal, J. C., Bates, P. D., Fewtrell, T. J., _et al._ (2009c). Distributed whole city water level measurements from the Carlisle 2005 urban flood event and comparison with hydraulic model simulations. *Journal of Hydrology*, **368**, 42–55.

Neal, J., Fewtrell, T., Bates, P. D., and Wright, N. (2010). A comparison of three parallelisation methods for 2D flood inundation models. *Environmental Modelling and Software*, **25**, 398–411.

Neal, J., Schumann, G., Fewtrell, T. J., *et al.* (2011). Evaluating a new LISFLOOD-FP formulation with the summer 2007 floods in Tewkesbury UK. *Journal of Flood Risk Management*, **4**, 88–95.

Néelz, S., and Pender, G. (2006). The influence of errors in Digital Terrain Models on flood flow routes. In *River Flow 2006*, Lisbon, Portugal: IAHR, 1955–1962.

Nelson, P. A., Smith, J. A., and Miller, A. J. (2009). Evolution of channel morphology and hydrologic response in an urbanizing drainage basin. *Earth Surface Processes and Landforms*, **31**, 1063–1079.

NERC (Natural Environment Research Council) (1975). *Flood Studies Report*. London.

Oberstadler, R., Hönsch, H., and Huth, D. (1997). Assessment of the mapping capabilities of ERS-1 SAR data for flood mapping: a case study in Germany. *Hydrological Processes*, **10**, 1415–1425.

Ohl, C., and Tapsell, S. (2000). Flooding and human health: the dangers posed are not always obvious. *British Medical Journal*, **321**(7270), 1167–1168.

Opperman, J. J., Galloway, G. E., Fargione, J., *et al.* (2009). Sustainable floodplains through large-scale reconnection to rivers. *Science*, **326**, 1487–1488.

Oreskes, N., Shraderfrechette, K., and Belitz, K. (1994). Verification, validation, and confirmation of numerical-models in the earth-sciences. *Science*, **263**(5147), 641–646.

Otsu, N. (1979). A threshold selection method from gray-level histograms. *IEEE Transactions on Systems, Man, and Cybernetics*, **9**, 62–66.

Padi, P., Di Baldassarre, G., and Castellarin, A. (2011). Floodplain management in Africa: large scale analysis of flood data. *Physics and Chemistry of the Earth*, **36**(7–8), 292–298.

Pappenberger, F., and Beven, K. J. (2006). Ignorance is bliss: or seven reasons not to use uncertainty analysis. *Water Resources Research*, **42**, W05302, doi:10.1029/2005WR004820.

Pappenberger, F., Beven, K., Horritt, M., and Blazkova, S. (2005). Uncertainty in the calibration of effective roughness parameters in HEC-RAS using inundation and downstream level observations. *Journal of Hydrology*, **302**, 46–69.

Pappenberger, F., Matgen, P., Beven, K. J., *et al.* (2006). Influence of uncertain boundary conditions and model structure on flood inundation predictions. *Advances in Water Resources*, **29**, 1430–1449.

Pappenberger, F., Beven, K. J., Frodsham, K., Romanovicz, R., and Matgen, P. (2007). Grasping the unavoidable subjectivity in calibration of flood inundation models: a vulnerability weighted approach. *Journal of Hydrology*, **333**(2–4), 275–287.

Pelletier, M. P. (1987). Uncertainties in the determination of river discharge: a literature review. *Canadian Journal of Civil Engineering*, **15**, 834–850.

Pender, G. (2006). Briefing: Introducing the Flood Risk Management Research Consortium. *Proceedings of the Institution of Civil Engineers: Water Management*, **159**, 1–6.

Pender, G., and Faulkner, H. (2010). *Flood Risk Science and Management*. Oxford, UK: Wiley-Blackwell.

Petersen-Øverleir, A. (2004). Accounting for heteroscedasticity in rating curve estimates. *Journal of Hydrology*, **292**, 173–181.

Pinder, G. E., and Sauer, S. P. (1971). Numerical simulation of flood wave modification due to bank storage effects. *Water Resources Research*, **7**, 63–70.

Ponce, V. M., and Simons, D. B. (1977). Shallow wave propagation in open channel flow. *Journal of the Hydraulic Division, Proceedings American Society of Civil Engineers*, **103**(12), 1461–1475.

Poole, G. C., Stanford, J. A., Frissell, C. A., and Running, S. W. (2002). Three-dimensional mapping of geomorphic controls on flood-plain hydrology and connectivity from aerial photos. *Geomorphology*, **48**(4), 329–347.

Preissmann, A. (1961). Propagation of translatory waves in channels and rivers. In *Proceedings of the 1st Congress de l'Association Francaise de Calcul*, Grenoble, France, 433–442.

Price, R. K. (1975). Flood routing studies. In *Flood Studies Report* 3, Natural Environment Research Council, London.

Raclot, D. (2006). Remote sensing of water levels on floodplains: a spatial approach guided by hydraulic functioning. *International Journal of Remote Sensing*, **27**(12), 2553–2574.

Rantz, S. E., *et al.* (1982). *Measurement and Computation of Streamflow*. US Geological Survey Water Supply Paper 2175, available at http://water.usgs. gov/pubs/wsp/wsp2175/.

Refsgaard, J. C. (2001). Towards a formal approach to calibration and validation of models using spatial data. In Grayson, R. B., and Blöschl, G. (eds.), *Spatial Patterns in Catchment Hydrology: Observations and Modelling*, Cambridge, UK: Cambridge University Press, 329–354.

Roca, M., and Davison, M. (2009). Two dimensional model analysis of flash-flood processes: application to the Boscastle event. *Journal of Flood Risk Management*, **3**(1), 63–71.

Romanowicz, R., and Beven, K. (1996). Bayesian calibration of flood inundation models. In Anderson, M. G., Walling, D. E., and Bates, P. D. (eds.), *Floodplain Processes*, Chichester, UK: John Wiley and Sons, 297–318.

Romanowicz, R., and Beven, K. (1998). Dynamic real-time prediction of flood inundation probabilities. *Hydrological Sciences Journal*, **43**(2), 181–196.

Romanowicz, R., and Beven, K. (2003). Estimation of flood inundation probabilities as conditioned on event inundation maps. *Water Resources Research*, **39**(3), 1073–1085.

Romanowicz, R., Beven, K. J., and Tawn, J. (1996). Bayesian calibration of flood inundation models. In Anderson, M. G., Walling, D. E., and Bates, P. D. (eds.), *Floodplain Processes*, Chichester, UK: John Wiley and Sons, 333–360.

Sagris, V., Petrov, L., and Lavalle, C. (2005). *Towards an Integrated Assessment of Climate Change-Induced Sea-Level Rise in the Baltic Sea: An Example for the City of Pärnu (Estonia)*. EC-Joint Research Centre, Ispra, Italy.

Samuels, P. G. (1990). Cross section location in one-dimensional models. In White, W. R. (ed.), *International Conference on River Flood Hydraulics*, Chichester, UK: John Wiley and Sons, 339–350.

Sanderson, E. W., Jaiteh, M., Levy, M. A., *et al.* (2002). The human footprint and the last of the wild. *BioScience*, **52**, 891–904.

Savant, G., Berger, C., McAlpin, T. O., and Tate, J. N. (2010). An implicit finite element hydrodynamic model for dam and levee breach. *Journal of Hydraulic Engineering*, doi:10.1061/(ASCE)HY.1943-7900.0000372.

Sayers, P. B., Hall, J. W., and Meadowcroft, I. C. (2002). Towards risk-based flood hazard management in the UK. *Proceedings of the Institution of Civil Engineers*, Paper 12803, **150**, 36–42.

Schumann, G., and Di Baldassarre, G. (2010). The direct use of radar satellites for event-specific flood risk mapping. *Remote Sensing Letters*, **1**(2), 75–84.

Schumann, G., Matgen, P., Hoffmann, L., *et al.* (2007a). Deriving distributed roughness values from satellite radar data for flood inundation modelling. *Journal of Hydrology*, **344**(1–2), 96–111.

Schumann, G., Hostache, R., Puech, C., *et al.* (2007b). High-resolution 3D flood information from radar imagery for flood hazard management. *IEEE Transactions on Geoscience and Remote Sensing*, **45**(6), 1715–1725.

Schumann, G., Matgen, P., Cutler, M. E. J., *et al.* (2008). Comparison of remotely sensed water stages from LiDAR, topographic contours and SRTM. *Journal of Photogrammetry and Remote Sensing*, **63**, 283–296.

Schumann, G., Di Baldassarre, G., and Bates, P. D. (2009a). The utility of space-borne radar to render flood inundation maps based on multi-algorithm ensembles. *IEEE Transactions on Geoscience and Remote Sensing*, **47**(8), 2801–2807.

Schumann, G., Bates, P. D., Horritt, M. S., Matgen, P., and Pappenberger, F. (2009b). Progress in integration of remote sensing-derived flood extent and stage data and hydraulic models. *Reviews of Geophysics*, **47**, RG4001, doi:10.1029/2008RG000274.

Schumann, G., Di Baldassarre, G., Alsdorf, D. E., and Bates, P. D. (2010). Near real-time flood wave approximation on large rivers from space: application to the River Po, Northern Italy. *Water Resources Research*, **46**, W05601.

Sellin, R. H. J. (1964). A laboratory investigation into the interaction between the flow in the channel of a river and that over its floodplain. *La Houille Blanche*, **7**, 793–801.

Sellin, R. H. J., and Willetts, B. B. (1996). Three-dimensional structures, memory and energy dissipation in meandering compound channel flow. In Anderson, M. G., Walling, D. E., and Bates, P. D. (eds.), *Floodplain Processes*, Chichester, UK: John Wiley and Sons, 255–298.

Simonovic, S. (2012). *Floods in a Changing Climate: Risk Management*. International Hydrology Series, Cambridge, UK: Cambridge University Press.

Smith, D. M. (1996). Speckle reduction and segmentation of synthetic aperture radar images. *International Journal of Remote Sensing*, **17**, 2043–2057.

Smith, K. (1996). *Environmental Hazards: Assessing Risk and Reducing Disaster*, 2nd edition. London: Routledge.

and Shrestha, D. L. (2008). A novel method to estimate ...y using machine learning techniques. *Water Resources* .., **45**, W00B11.

.ooshian, S., Gupta, V. K., and Fulton, J. L. (1983). Evaluation of maximum likelihood parameter estimation techniques for conceptual rainfall-runoff models: influence of calibration data variability and length on model credibility. *Water Resources Research*, **19**(1), 251–259, doi:10.1029/WR019i001p00251.

Staatscourant (1998). *Regeling oogstschade 1998*. The Hague, the Netherlands: Ministerie van Algemene Zaken, 16–17.

Stelling, G. S., and Duynmeijer, S. P. A. (2003). A staggered conservative scheme for every Froude number in rapidly varied shallow water flows. *International Journal for Numerical Methods in Fluids*, **43**, 1329–1354.

Strelkof, T., and Katapodes, N. D. (1977). Border irrigation hydraulics with zero inertia. *Journal of Irrigation and Drainage Engineering*, **103**, 325–342.

Swanson, K. M., Watson, E., Aalto, R., *et al.* (2008). Sediment load and floodplain deposition rates: comparison of the Fly and Strickland rivers, Papua New Guinea. *Journal of Geophysical Research*, **113**, F01S03, doi:10.1029/2006JF000623.

Szöllösi-Nagy, A. (2009). Learn from your errors – if you can! Reflections on the value of hydrological forecasting models. Inaugural address, UNESCO-IHE, Delft, the Netherlands.

Szöllösi-Nagy, A., and Mekis, E. (1988). Comparative analysis of three recursive real-time river flow forecasting models: deterministic, stochastic, and coupled deterministic-stochastic. *Stochastic Hydrology and Hydraulics*, **2**, 17–33.

Teegavarapu, R. (2012). *Floods in a Changing Climate: Extreme Precipitation*. International Hydrology Series, Cambridge, UK: Cambridge University Press.

Todini, E., and Bossi, A. (1986). PAB (Parabolic and Backwater) an unconditionally stable flood routing scheme particularly suited for real time forecasting and control. *Journal of Hydraulic Research*, **24**(5), 405–424.

Toro, E. (1997). *Riemann Solvers and Numerical Methods for Fluid Dynamics*. Berlin, Germany: Springer.

Uhlenbrook, S., Di Baldassarre, G., Bhattacharya, B., *et al.* (2011). Flood management in a changing world: why and how do we have to change our approach in education? *5th International Conference on Flood Management* (ICFM5), 27–29 September 2011, Tsukuba-Japan.

UNEP/GRID (2000). Potential impact of sea level rise: Nile Delta. In UNEP/GRID-Arendal Maps and Graphics Library. Retrieved from http://maps.grida.no/go/graphic/potential-impact-of-sea-level-rise-nile-delta. Last accessed November 1, 2011.

UN-ISDR (2004). *Terminology: Basic Terms of Disaster Risk Reduction*. UN International Strategy for Disaster Reduction, www.unisdr.org/eng/library/lib-terminology-eng%20home.htm.

UN-ISDR (2009). *Reducing Disaster Risks Through Science: Issues and Actions*. Full report of the ISDR Scientific and Technical Committee.

van Leer, B. (1979). Towards the ultimate conservative difference scheme V. A second order sequel to Godunov's method. *Journal of Computational Physics*, **32**, 101–136.

Van Manen, S. E., and Brinkhuis, M. (2005). Quantitative flood risk assessment for polders. *Reliability Engineering and System Safety*, **90**, 229–237.

Viglione, A., Laio, F., and Claps, P. (2007). A comparison of homogeneity tests for regional frequency analysis. *Water Resources Research*, **43**, W03428, doi:10.1029/2006WR005095.

Villanueva, I., and Wright, N. G. (2006). Linking Riemann and storage cell models for flood prediction. *Proceedings of the Institution of Civil Engineers: Water Management*, **159**, 27–33.

Vis, M., Klijn, F., De Bruijn, K. M., and Van Buuren, M. (2003). Resilience strategies for flood risk management in the Netherlands. *International Journal of River Basin Management*, **1**(1), 33–40.

Vrisou van Eck, N., and Kok, M. (2001). *Standaardmethode Schade en Slachtoffers als gevolg van overstromingen*. Dienst Weg- en Waterbouwkunde, Ministerie van Rijkswaterstaat, the Netherlands, Publication W-DWW-2001–028.

Vrugt, J. A., ter Braak, C. J. F., Gupta, H. V., and Robinson, B. A. (2009). Equifinality of formal (DREAM) and informal (GLUE) Bayesian approaches in hydrologic modeling? *Stochastic Environmental Research and Risk Assessment*, **23**(7), 1011–1026.

Wagener, T., Sivapalan, M., Troch, P. A., *et al.* (2010). The future of hydrology: an evolving science for a changing world. *Water Resources Research*, **46**, W05301.

Wagenmakers, E. J. (2003). How many parameters does it take to fit an elephant? *Journal of Mathematical Psychology*, **47**, 580–586.

Ward, R. C. (1978). *Floods: A Geographical Perspective*. London: Macmillan.

Whitham, G. (1974). *Linear and Nonlinear Waves*. New York: John Wiley and Sons.

Wilby, R. L., Beven, K. J., and Reynard, N. S. (2008). Climate change and fluvial flood risk: more of the same? *Hydrological Processes*, **22**, 2511–2523, doi: 10.1002/hyp.6847.

Wilson, C. A. M. E., Stoesser, T., and Bates, P. D. (2005). Modelling of open channel flow through vegetation. In Bates, P. D., Lane, S. N., and Ferguson, R. I. (eds.), *Computational Fluid Dynamics: Applications in Environmental Hydraulics*, Chichester, UK: John Wiley and Sons.

Woodhead, S., Asselman, N., Zech, Y., *et al.* (2009). *Evaluation of Inundation Models*. FP6 Integrated Project FLOODsite, Public Report, www.floodsite.net, HR Wallingford, UK.

World Meteorological Organisation (1994). *Guide to Hydrological Practice*, WMO Publication No. 168.

Wright, N. G., Asce, M., Villanueva, I., *et al.* (2008). Case study of the use of remotely sensed data for modeling flood inundation on the River Severn, UK. *Journal of Hydraulic Engineering*, **134**(5), 533–540.

Yamazaki, D., Kanae, S., Kim, H., and Oki, T. (2011). A physically based description of floodplain inundation dynamics in a global river routing model. *Water Resources Research*, **47**, W04501.

Yu, D., and Lane, S. N. (2006). Urban fluvial flood modelling using a two-dimensional diffusion-wave treatment. Part 1: Mesh resolution effects. *Hydrological Processes*, **20**, 1541–1565.

Zwenzner, H., and Voigt, S. (2009). Improved estimation of flood parameters by combining space based SAR data with very high resolution digital elevation data. *Hydrology and Earth System Sciences*, **13**, 567–576.

Index

5

Printed in the United States
by Baker & Taylor Publisher Services